Integrated agriculture-aquaculture

A primer

FAO FISHERIES TECHNICAL PAPER
407

INTERNATIONAL INSTITUTE OF RURAL RECONSTRUCTION

ICLARM
THE WORLD FISH CENTER

Food and Agriculture Organization of the United Nations

Rome, 2001

2005
Published by Arrangement with the
Food and Agriculture Organization
of the United Nations
by
DAYA PUBLISHING HOUSE
Delhi - 110 035
for Sale and Distribution only in South-East Asia

"This book was originally published by the Food and Agriculture Organization of the United Nations (FAO) as FAO Fisheries Technical Paper No. 407."

"The designations employed and the presentation of material in this publication do not imply the expression of any opinion whatsoever on the part of the Food and Agriculture Organization of the United Nations concerning the legal status of any country, territory, city or area of its authorities, or concerning the delimitation of its frontiers or boundaries."

"The designations "developed" and "developing" economies are intended for statistical convenience and do not necessarily express a judgement about the stage reached by a particular country, territory or area in the development process."

First Indian Edition 2005

ISBN 81-7035-406-4

Published by : **Daya Publishing House**
1123/74, Deva Ram Park
Tri Nagar, Delhi - 110 035
Phone: 27383999
Fax: (011) 23244987
e-mail : dayabooks@vsnl.com
website : www.dayabooks.com

Showroom : 4762-63/23, Ansari Road, Darya Ganj,
New Delhi - 110 002
Phone: 23245578, 23244987

Printed at : **Chawla Offset Printers**
Delhi - 110 052

PRINTED IN INDIA

Preparation of This Document

This document contains an edited and slightly revised version of a previously published integrated agriculture-aquaculture (IAA) technology information kit[1]. In response to expressed need for wider distribution, but limited copies of the original document, this reprint was produced. The activity was initiated and supported by an FAO team led by Dr. Ziad Shehadeh, and seen through to finalization in the framework of the integrated agriculture-aquaculture activities of Dr. Matthias Halwart of the Fishery Resources Division of the Food and Agriculture Organization of the United Nations.

The original version of the information kit was externally reviewed by four IAA experts: Mr. Weimin Miao (Wuxi, People's Republic of China), Dr. Le Thanh Luu (Hanoi, Viet Nam), Dr. Dilip Kumar (Bangkok, Thailand) and Dr. David C. Little (Stirling, Scotland). Their reviews, together with comments and information contributed by other experts, were synthesized and implemented where possible and of use, given the objectives of reprinting and the resources available.

In October 2000, a publications team to finalize the document was assembled on the campus of IIRR in Cavite, Philippines, consisting of Ms. Marie Sol Sadorra-Colocado (Editor), Ms. Florisa Norina Luna-Carada (Senior DTP/Layout), Mr. Ivan Roy Mallari (Junior DTP/Layout), Mr. Rizalino S. Bautista (Junior DTP/Layout), Ms. Ely G. Lumdang (Proofreader/Encoder), Mr. Ricardo E. Cantada (Artist/Draftsperson) and Ms. Lilibeth V. Mercado (Encoder/Publications Clerk). The team was supervised by Dr. Mark Prein (ICLARM), Dr. Matthias Halwart (FAO) and Dr. Julian Gonsalves (IIRR).

The FAO editorial style was implemented. Where applicable, figures and tables were revised, redrawn or modified, and in few cases omitted to improve clarity of presentation. As far as these were known, errors were corrected. In contrast to the first version in 1992, this version was prepared for printing fully electronically.

The entire document is intended to be available for downloading from the FAO fisheries website (http://www.fao.org/fi).

> The *cover diagram* shows a hypothetical rural smallholder farm with multiple enterprises and examples of bioresource-flows, i.e. intentionally farmer-managed and farmer-moved plant and animal wastes from enterprises, which act as nutrient inputs to other enterprises, thereby enhancing their productivity. Wastes can be from on-farm and off-farm sources. Aside from waste recycling, non-waste off-farm nutrient inputs such as fertilizers and feeds can play a role, particularly when production capacity is to be increased. For the sake of visual presentability, not all possible flows are shown, as the decision about which wastes to recycle and to which enterprise these should be added, is up to the farmers based on their agroecological and socioeconomic situation. Therefore, only some examples of within-farm nutrient flows are shown, excluding those of off-farm nutrient inputs, flows of wastes leaving the farm, and flows of products to the household for home consumption and to off-farm markets (drawing by Ricardo E. Cantada, designed by Mark Prein).

[1]IIRR and ICLARM (eds.). 1992. Farmer-proven integrated agriculture-aquaculture: a technology information kit. International Institute of Rural Reconstruction, Silang, Cavite, Philippines, and International Center for Living Aquatic Resources Management, Makati, Philippines. 183 p.

FAO/ICLARM/IIRR.
Integrated agriculture-aquaculture: a primer.
FAO Fisheries Technical Paper. No. 407. Rome, FAO. 2001. 149p.

Abstract

This document is an edited and slightly revised version of a previously published integrated agriculture-aquaculture (IAA) technology information kit. It contains 38 contributions in seven sections, outlining the basic issues and characteristics of IAA systems and making generous use of pictorial drawings and visual representations.

Sociocultural, economic and environmental considerations in introducing IAA technologies are presented in the first four contributions. This section is followed by an overview of integrated farming systems, with six examples provided, ranging from integrated grass-fish and embankment-fish systems in the People's Republic of China, over the VAC system in northern Viet Nam to short-cycle methods in seasonal ponds and ditches in Bangladesh. The next section has four papers dealing with livestock-fish integration of chicken-, duck- and pig-based systems. Two sections with a total of 16 presentations tackle several aspects of rice-fish systems, starting with eight technical examples from five countries, including irrigation systems, and in coastal areas with shrimp and in freshwater areas with prawn. Eight more presentations give recommendations on site selection, ricefield preparation, fish stocking, feeding, rice management and integrated pest management issues within rice-fish culture. Another section with four papers deals with aspects of fish feeding and management in IAA, such as the use of animal manures, domestic sewage and biogas slurry in ponds, as well as plant sources as fish feed. The last section contains four contributions on fish breeding and nursing, focusing on fry and fingerling production and emphasizing carp species. This includes a description of carp spawning in wheat fields and fry nursing in ricefields as off-season activities, as well as fry-to-fingerling rearing in ricefields.

This primer aims to give decisionmakers in governmental and non-governmental organizations and in other organizations concerned with agriculture and rural development an overview and a basis for understanding the principles of IAA, and to help them decide whether to embark on IAA activities and include these in their program portfolio. For those who work directly with farmers, this primer aims at providing good examples of IAA, but it is *not* intended to be a compilation of procedures that should be strictly followed. Rather, this primer should help convince its readers/users that farmers can improve their livelihoods by either introducing IAA, or by further developing and improving the many IAA opportunities on their existing farms within their communities.

Distribution:

FAO Fisheries Department
FAO Fishery Regional and Sub-regional Officers
Authors

Contents

Preface

Starting in the late 1980s, the International Institute of Rural Reconstruction (IIRR) began to convene workshops for the purpose of documenting exemplary practices in sustainable agriculture. The resulting publication usually was a user-friendly and highly illustrated source book of ideas targeted to development workers and trainers.

Resource persons are invited to such workshops (also known as "writeshops") to present their ideas and experiences in the form of short papers, which are then subjected to critical review by fellow participants. Communication specialists and design/desktop publishing staff assist in print presentation. The revised materials are reviewed again until all changes are acceptable. Only then are these generated materials considered suitable and relevant for immediate dissemination and use.

What is unique about this process is that these materials are generated and developed by bringing scientists, development workers and communication specialists to a workshop specifically for such purpose.

IIRR and the International Center for Living Aquatic Resources Management (ICLARM) valued the idea of developing a publication on integrated agriculture-aquaculture to help improve the quality of life of farmers on smallholdings in Asia. The two institutions, supported by the Netherlands Organization for International Development and Cooperation (NOVIB) and the Association of Southeast Asian Nations (ASEAN)-Canada Fund, organized and conducted a workshop at IIRR in Cavite, Philippines, in February 1992. This resulted in the publication of the "Farmer-proven integrated agriculture-aquaculture: a technology information kit" which was deliberately made copyright-free to allow for reprinting and wider distribution, provided the source was cited.

The 2000 printed copies were distributed to extensionists, farmers, university students, scientists and decisionmakers in governmental, nongovernmental and local organizations, and bilateral aid donors. Feedback from users revealed that the kit was used in training courses and communications such as posters and lectures. It was highly sought after, so that more sets had to be photocopied, and reprinting became an issue.

The Food and Agriculture Organization of the United Nations (FAO), which has a long history of collaborating with IIRR and ICLARM, considered reprinting the kit an important activity to complement current efforts of its Inland Water Resources and Aquaculture Service to build awareness among policymakers that aquaculture has an important role to play in poor peoples' livelihoods[1] and to document successful cases of small-scale aquaculture in different environments[2]. In the context of the Organization's efforts to help member countries achieve food security and alleviate poverty, the kit was considered to be very valuable and useful, and a powerful communication tool with potential for wider application in many countries, particularly through FAO's Partnership Programmes and the Special Programme for Food Security (SPFS). FAO therefore teamed up with IIRR and ICLARM in a joint effort to edit and revise the original publication and publish it as a primer on IAA in the FAO Fisheries Technical Paper series.

This primer aims to give decisionmakers in governmental and non-governmental organizations and in other organizations concerned with agriculture and rural development an overview and a basis for understanding the principles of IAA, and to help them

---

[1]FAO. 2000. Small ponds make a big difference. FAO, Rome, Italy, 30 p.

[2]IIRR, IDRC, FAO, NACA and ICLARM (in press). Utilizing Different Aquatic Resources for Livelihoods in Asia: A Resource Book. Proceedings of a Workshop, 18-28 September 2000. International Institute of Rural Reconstruction, Silang, Cavite, Philippines. 407 p.

decide whether to embark on IAA activities and include these in their program portfolio. The target beneficiary group of this publication are the small-scale farmers who already have a small aquaculture activity (e.g. a small pond or a rice-fish system) and could benefit from improved systems as shown in this publication, and the farmers without any form of aquaculture on their farms but with access to appropriate sites and resources to establish an aquaculture component as a means of diversification. For the latter, a simple and low-cost starting point is the use of existing, otherwise unutilized on-farm, or easily accessible off-farm, resources such as wastes as a means to fertilize their ponds. This integration can take on a multitude of forms, many of which are described in the presentations in this book. The possible forms of farm integration are limited mainly by the resources available to the farmers, and their creativity.

The IAA operations usually take up a minor area on the farm, compared to major activities such as staple food crops, cash crops and orchards. Yet these operations can be very important and highly productive components, when efficiency is regarded on a value-per-area basis. Environmental and agroecosystem characteristics should be supportive of all components of the integrated system for this to function optimally and most beneficially for the farmer.

The previous approach of introducing stand-alone fish culture enterprises was often not successful for novices and has led to countless failures in small-scale aquaculture development. Instead, IAA has proven to be a viable entry point into fish farming, which the farmer can later improve with increased expertise and specialization. It is not the aim of this publication to convince traditional, small-scale farmers to abandon their ongoing farming activities and abruptly become fish farmers as an exclusive occupation. IAA relies on linkages and synergies among different on-farm and off-farm enterprises. It aims to encourage farmers towards diversification and intensification, but without negative effects of overuse of external inputs and monocultures.

The descriptions in the presentations on the calendars and schedules of activities refer to the specific location and time (i.e. early 1990s) in which the case descriptions were written, often with a reference to the countries where the method was developed or is being applied. Situations and agroecological context will differ and vary in other locations with different seasons. The reader should be encouraged to examine carefully the local context of the area in which the application of IAA is intended. The IAA systems described are from a range of applications - experimental, researcher-managed on-farm trials, with downscaled quantities and dimensions from commercial systems, descriptions of large-scale systems with some applications in small-scale systems, all the way to farmer-developed and widely implemented systems.

This publication is not a compilation of procedures that should be strictly followed. Rather, this primer should help convince its readers/users that farmers can discover and develop opportunities for IAA activities on their existing farms within their communities. Readers are encouraged to note that it is the idea and principle of IAA, and not the individual examples and details of the descriptions, which should be absorbed and later applied. Farmers should use the given proportions of component size, types and amounts of material flows, fish and plant stocking densities only as guide on which to base their own trials.

Original contributions were edited and revised. Importantly, boxes were provided at the end of most presentations with a summary of reviewers' and editors' comments, which are intended to give updated views on the topics and further background information for application.

The bibliography and the designations and affiliations of participants were as in the original publication.

In terms of appearance, readers should bear in mind that the present publication has used modern DTP tools but relied on eight year-old hand-drawn and written pictures (although a number of them were redrawn for this reprint), legends and captions, originating from the 1992 version which was as a loose-leaf collection of information sheets meant to be copied as handouts. Figures and tables were updated to meet the general objectives of this reprint, albeit with the aim for cost effectiveness, and adapted to the FAO editorial style for their Fisheries Technical Paper series.

Mentions of pesticide brands and types which may be outdated or not appropriate for a specific use are made because they were in use at the particular time and place. Their mention does not in any way represent an endorsement by FAO, IIRR or ICLARM.

It is intended that this primer will be downloadable from the FAO website (http:/www.fao.org/fi).

Rome, December 2000

Matthias Halwart
Fishery Resources Officer (Aquaculture)
Inland Water Resources and Aquaculture Service
FAO

Julian Gonsalves
Vice President – Programs
IIRR

Mark Prein
Senior Scientist/Leader
Freshwater Resources Research Program
ICLARM

Introduction

Without exception, Asian policymakers and planners face a crisis of continuing rural poverty. Each year, millions more children are added to the farming households without much hope of a better livelihood. Each year, millions of hectares of farming's natural resource base become further degraded. Modern farming methods with their high-external inputs and economies of scale may promise food but at the cost of pollution, marginalization of the poor, and fewer and fewer jobs. Somehow, small-scale farming systems must provide a reasonable rural livelihood, a clean conserved environment, and adequate food, fuel and fibre products.

No doubt, new policies will be needed to protect and foster such development. No doubt, new institutions for marketing, banking and education will be needed at community, local and national levels. No doubt, a higher level of farming management and professionalism will be required. But all of these require governments to be serious and imaginative about rural development.

One option for sustainable development in farming is small-scale integrated agriculture-aqua- culture. The diversification that comes from integrating crops, vegetables, livestock, trees and fish imparts stability in production, efficiency in resource use and conservation of the environment. Uncertainty in markets and climate is countered by a wide array of enterprises. In integrated farming, wastes of one enterprise become inputs to another and, thus, optimize the use of resources and lessen pollution. Stability in many contrasting habitats permits diversity of genetic resources and survival of beneficial insects and other wildlife. Integrated agriculture-aquaculture offers special advantages over and above its role in waste recycling and its importance in encouraging better water management for agriculture and forestry. Fish are efficient converters of low-grade feed and wastes into high-value protein. Fish are the greatest sources of animal protein in rural Asia. For rural households, fish are small units of cash or food which can be harvested more or less at will without loss of weight or condition. While these systems are labour-intensive, they do save labour from fetching water, gathering wood and forage, and fishing in nearby rivers and streams. All of these are elaborated in this technology information kit.

The many examples of integrated farming systems from around Asia that are presented in this technology information kit are not given as models to copy or emulate exactly. Rarely can such complex systems be built from scratch. Indeed, many of the technical and budget details will not apply in every case. Rather, examples are given to show what is possible and to stimulate a process of integration on the farm. What other farmers have proven is shared here to help people who work directly with farmers to facilitate the addition of new resource flows, the integration of new enterprises, the substitution of external inputs and the rehabilitation of degraded agroecosystems.

The kit suggests a procedure for evolving farming systems that share the characteristics described herein. Moreover, we have seen that this procedure not only captures the many levels of integration within a single group of households, but also stimulates households to increase levels of integration.

This technology information kit seeks to stimulate people who work directly with farmers to develop small-scale farms that provide a reasonable rural livelihood, a clean conserved environment, and food, fuel and fibre products.

Clive Lightfoot (ICLARM)
Julian Gonsalves (IIRR)
1992, Philippines

Considerations in Introducing Integrated Agriculture-Aquaculture Technology

Sociocultural considerations when introducing a new integrated agriculture-aquaculture technology

Eric Worby

It is important to know how farmers understand the world before trying to introduce new technological options. Discover whether or not the new system can fit in well with the farmer's concerns, beliefs and values. Remember farmers are "scientists", too. They have been developing, testing and adopting their own technologies for centuries in ways that are tailored to their cultural setting. If you first make effort to learn from them about the fit between cultural outlook and technology, then you will have a much better idea of which new technologies they are likely to take an interest in.

Some general considerations

1. Even science is cultural. It is a belief system that incorporates certain values and goals, and promotes a particular view of the world.
   - Agricultural scientists and economists value precision in measurement and replicability of results, as well as maximizing efficiency and profitability.
   - Farmers may be motivated by goals and values that are different from those of scientists and economists.
   - Farmers may value security of livelihood for themselves and their children in the short and long term. They place a higher priority on preserving harmony in the community than on maximizing individual gain; or they may seek to acquire merit in the afterlife by contributing fish to a temple rather than selling them for money.
2. Cultural rules often limit what particular members of a given society (e.g. women versus men) can do. Cultural factors may determine who usually makes decisions, who is allowed to work in the fields, who may go to town to market produce and who may travel to a research station to attend demonstrations. These factors may set limits on the flexibility of households and communities to adopt new technologies. For example:
   - Women may not be permitted to catch fish, but they may be the ones who sell them.

   Considerations of gender, religious beliefs, caste or clan membership may limit the distribution of benefits to be derived from farming innovations.
3. Interactions between extension agents or institutions and farmers may be constrained by culture.
   - It may be unacceptable for male extension workers to speak freely with women. Or a young extension worker may feel uncomfortable giving instructions to a distinguished community elder.

4. Culture changes over time. Children often have different beliefs, attitudes and values than their parents. This can cause conflicts over resource use priorities. For example:
   - Children may aspire to enter nonfarm occupations or may be less concerned with respecting religious taboos.
5. Communities and consumption need to be considered. Farming communities are often divided by factors, such as religion, caste, economic class and political affiliation. A given technology may not be suitable for the whole community and may increase conflict within it.

Consumption constraints

There is no reason to encourage people to raise fish if they will not eat the fish themselves and if they cannot find anyone who will buy the fish. The same is true for any livestock or vegetable product that may be part of an integrated farming technology. It is, therefore, essential to consider the local cultural and economic constraints on consumption before attempting to introduce such a new technology.

Cultural constraints on consumption may include:

1. Religious beliefs
   For example:
   - Muslims will not eat pig meat; many will not consume shellfish, but this depends on local custom and preferences.
   - Most Hindus will refuse cow meat; some castes eat no meat, fish or livestock product of any kind. Again, this varies among regions.
   - Some Buddhists will not kill and consume domesticated animals (including farmed fish) although they will eat wild fish.
2. Totemic beliefs
   - Especially in Africa, but also among tribal peoples in Asia, Melanesia and the Americas, some people are forbidden to eat the animal that stands for their clan group.
3. Beliefs about gender differences
   - In some societies, men may be allowed to eat certain foods that are forbidden to women and vice versa. Often, men expect to be given the most nutritious and preferred foods first. These factors may reduce the nutritional benefits that women receive from fish or livestock production. On the other hand, sometimes women can demand these foods when they are pregnant or nursing.
4. Beliefs about food cleanliness and health
   - Sometimes people believe that certain foods are unclean or will make them sick. For example, many people refuse to eat fish raised on animal excreta for these reasons.

Below is a consumption checklist to help you think about how cultural beliefs might affect the adoption of the new technology you want to introduce. What other technology might be more culturally appropriate?

This checklist will help you decide whether or not a new technology will generate products that will be available and acceptable to all members of the producing households as well as to buyers in the market. However, you must still make a separate assessment of the long-term level of demand and prices in the markets of a farmer's product

Consumption checklist

Will be available and acceptable to:	Inputs		Outputs					
	Manure	Others	Fish	Pig meat	Cow meat	Poultry meat	Eggs/ milk	Others
Women								
Pregnant/nursing women								
Children								
Men								
Religious/totemic groups								
Local markets								
Distant markets								

before deciding whether or not a given technology will be viable. (See this volume.)

Labour time

In most farming communities, women and men do different kinds of tasks on and off the farm and in the house. A new integrated farming system technology will usually require changes in the way members of the farming household use their time. Some might have a greater burden of work (e.g. feeding fish or livestock, repairing dikes, selling fish) and need to reduce the time they spend on other activities. But this is not always true. Sometimes, new tasks can be easily combined with present activities (e.g. digging a trench can supply fertilizer for horticulture crops on an embankment) or children and elders can perform new tasks that are not physically demanding but costly in terms of time (e.g. feeding fish in a distant pond).

The following labour requirement checklist will help you think about these problems and whether or not they can be easily solved by the farming household. But remember, households differ. Some have many young children who require supervision. Sometimes, an elderly widow lives alone and does most things by herself because her children have gone to find jobs in town. How can an integrated system help someone like her to increase her food and income without demanding more labour time? Are there neighbours, relatives or a women's group with whom she can cooperate and get help from?

For each task on the checklist, make a mark under "Present" if the category of family member (children, women, men, elders) contributes substantial labour under the existing system. Then make a mark under "Future" if they will need to contribute once the newly integrated system is adopted.

Labour requirement checklist

	Children		Women		Men		Elders	
	Present	Future	Present	Future	Present	Future	Present	Future
Field labour								
- Field								
- Preparation								
- Pesticide and fertilizer								
- Weeding								
- Harvesting								
Postharvest								
Processing								
- Grain								
- Livestock								
- Fish								
Livestock management								
- Feeding								
- Milking, collecting manure								
- Eggs								
- Pen maintenance								
- Herding								
Housework								
- Cooking/cleaning								
- Building/ maintenance								
Child care								
Toolmaking/repair								
Marketing produce/ buying inputs								
Working for wages								
Others (trading, crafts, etc.)								

Decisionmaking in the household

Before introducing a new integrated agriculture-aquaculture (IAA) technology, it is important to consider who will make the management decisions that are crucial for its success. For example, elders might have ultimate authority in the household concerning when to sell crops or livestock, but make few day-to-day decisions on stocking rates, feeding and fertilizing.

Women often manage family finances, as well as making day-to-day decisions concerning food purchases and preparation. Because women are usually responsible for ensuring adequate nutrition for themselves and their children, they are often more motivated than men to adopt new technologies that provide nutritional benefits, such as fish culture. Also, women are eager to invest their time in improving the productivity of a resource over which they have control of both management and the harvested product (for example, a backyard pond).

Distribution of resources

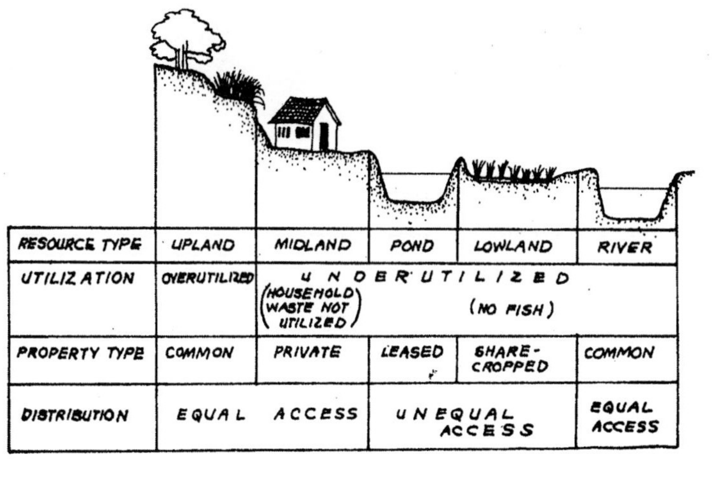

Distribution of resources

When we speak of "distribution", we mean the ways in which the resources needed for an integrated farming technology are made available to farmers. Some resources will be available on the farm and cost nothing (if the farm household owns them). These may have to be diverted from other uses, however, thus, constituting a hidden cost. Other resources may have to be borrowed, leased or purchased.

Before attempting to implement any of the technologies in this primer, you should try to answer the following questions together with the farmers you are working with. (You can do this as part of the drawing exercise discussed in the paper on working with new entrants to integrated agriculture-aquaculture, this volume.)

1. What resources are easily available on most farms in the area? (A new system should not depend upon resources that are scarce, difficult or expensive to obtain.)
2. Which of these resources is underutilized/not utilized? (A new system should focus on bringing these into the system.)
3. Which of these resources is overutilized/not utilized in a sustainable fashion? (A new system should strive toward restoring sustainability.)
4. Which of these resources is a common property resource? (A common property resource is one that is jointly drawn upon and managed by a community or a part of it, e.g. grazing lands,

On-farm resource availability and utilization checklist

	On-farm	Under-utilized	Must be diverted	Common property	Unequally distributed	Over-utilized
a. Land (proper soil slop, drainage)						
b. Water source (reliable, sufficient quality)						
c. Animal manure						
d. Green manure						
e. Household wastes (ash, sewage, food wastes)						
f. Grain processing by products						
g. Planting seed						
h. Fish fry-fingerlings						
i. Fish traps/nets						
j. Food processing tools						
k. Labour (knowledge/ skills, numbers, strength, timely availability)						

ponds, irrigation water, forest products. A new system should enhance the benefits all users receive from such resources.)

5. Which of these resources are controlled by only a small percentage of farmers or nonfarmers? (Farmers will be reluctant to invest in a system that requires resources that are not under their ownership or control, such as land that might be sold or an irrigation water supply that might be cut off someday.)

Managing risk: investing in social relationships

It is useful to remember that most farmers in the world have little margin for taking risks. Sometimes building a store of value to provide insurance against catastrophes (such as drought, flood, political upheaval, market instability, social and legal obligations) may be perceived by the farmer to be more desirable than investing for maximum returns.

Farmers view their ties with friends, neighbours and kinfolk as insurance against risk as well, since they will rely on them for help if disaster strikes. This is why farmers invest in social relationships—by sharing resources (such as money, tools and labour), paying visits, attending community celebrations and religious ceremonies and exchanging gifts. If a farmer harvests fish or poultry before these mature, it may be because he or she must meet a social obligation that can't be put off until later. Farmers should not be expected to make decisions in accordance with fixed models. Rather, the models for integrated technologies should be flexible enough to accommodate farmers' varying needs and perceptions of acceptable risk.

Most farm households will be familiar with the benefits that may be derived from integration in terms of reducing risk. Most likely, they already combine diverse enterprises (e.g. livestock, crops, wage labour, gardening) in order to protect themselves from possible failure of any single endeavour. The integration of agricultural enterprises with fish culture can increase household security by providing additional sources of income, improving cash flows over time and enhancing the long-term sustainability of the household and community resource base. Also, when nutrition is improved through integration, people become less vulnerable to illness.

Inequality between households

The farming households in any community are likely to have unequal access to resources and unequal control over their use. Often, extension agents focus on "leading or progressive farmers"—those with the greatest access to resources on the farm or with sufficient income to purchase these off the farm. Extension agents do this because it is easier to show a complete, complex system on a single farm or because the farmers often have more education and are more likely to think like the extension agent. These farms are often used to "demonstrate" the gains to be achieved from an integrated system. However, there are good reasons *not* to focus on resource-rich farmers in technology extension efforts.

- Resource-poor farmers will usually be discouraged from adopting a new technology they are shown on a wealthier farmer's farm. (They will think, "how can we possibly do it without land and cash?")
- Resource-rich farmers often control the distribution of inputs to poor farmers. Helping the rich farmers to expand may reduce access to resources by poor farmers, making it yet more difficult for them to adopt a new system that may improve their livelihood.
- When resource-poor farmers lose access to the means of survival, they are pressured to use the most fragile parts of the local ecosystem to gain a livelihood, often leading to environmental degradation. New technologies should be focused on solving this predicament; these should enable farming communities to manage environmental resources in a sustainable fashion by improving the security of livelihood for the community's poorest members.
- Remember, farmers continue to live in communities after outside advisers leave. Thus, it is a good idea to use extension agents who have intimate knowledge of the community they serve and to involve the whole community in choosing new and locally appropriate

systems. If one farmer shows a rapid increase in visible wealth after adopting a new technology, others may sabotage his investments, or feel envious and isolate themselves from the community.

Integration can reduce inequality in communities if the primary beneficiaries are the resource-poor community members.

- By making the resources they have access to more productive, poor farmers become less dependent on loans or favours from wealthy farmers.
- Involving resource-poor farmers in designing new integrated technologies may be a way of strengthening their control over their own lives and giving them better organizational capacity and power in the community.

Issues for further consideration

Through adoption of integrated farming, farmers can develop an improved understanding of resource use. How does this insight lead to further application of this knowledge elsewhere in people's livelihoods? Seasonality is an important aspect in fish farming and influences livelihood options in general.

The relationship between extension agents and farmers as described above is, unfortunately, not the norm. In reality, in most parts of developing countries, farmers have never seen a government extension agent. Considering this, alternative methods (which exist) for assessing and disseminating information may be sought.

The considerable variability in characteristics of families and communities, such as living individually or in joint households, levels of literacy and education, existing farming activities, food preferences, faiths and taboos make it difficult for occasional outside visitors, such as extension agents, to easily suggest appropriate technologies to adopt. After community-level discussions and presentations of a portfolio of options in adequate simplicity and format, individual families can decide to seek further advice on technologies they consider appropriate to their specific situation.

Benefits which nonproducers can gain from integrated farming in any area are potential employment and greater access to cheaper, nutritious food. Where fish culture cannot be adopted by the poorest sector, the latter may still be involved and/or benefit.

Economic considerations in introducing integrated agriculture-aquaculture technology

Mahfuzuddin Ahmed and Mary Ann P. Bimbao

How to make your farm budget

First, make a *cost sheet.*

- List the things that are required for you to use the technology.
- Write down how much is needed, its price and the amount paid.
- Add all amounts paid to find out the total costs.

Second, make an *income sheet.*

- List all the products from the technology that were sold.
- Write down how much is sold, at what price and the amount received.
- Add all amounts received to find out the total income.

Third, work out your *balance or profit sheet.*

- Write down the total income received from the technology.
- Write down the total costs that were required in doing the technology.
- Subtract the total amount paid for you to use the technology from the total amount received from the sales of the technology.

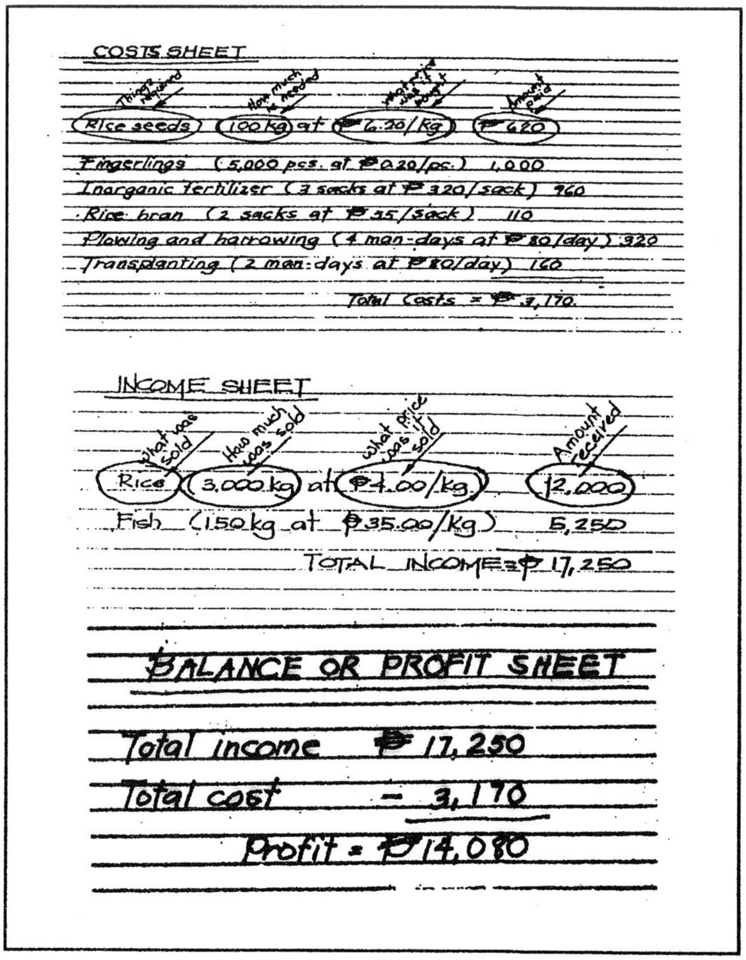

How to make your monthly cash flow

Cash outflow

- Work out your cash outflows. Note down the activities of the technology that required money and the cost involved and write these on the lower part of the calendar.
- Under January, the first month

of the technology, record plowing and harrowing where hired labourers were paid P320 for four days. Record also the purchase of rice seeds, which cost P620.

- For the second month of the technology, record transplanting activities, which required P160 as wage for hired labourers. Also, record the purchases and money paid for fingerlings, rice bran and inorganic fertilizer.
- Repeat recording on the calendar the activities of the technology that needed money and the amount paid for the succeeding months.

Cash inflow

- Work out your cash inflows. Note the products sold and the money received from these sales and write these on the upper part of the calendar.
- In April, the fourth month of the technology, 25 kg of fish were sold for P875.
- In May, 3 000 kg of rice were sold for P12 000 and 100 kg of fish for P3 500.
- Smaller fish were kept in the pond for further growout. A total fish harvest was done in June and 25 kg of fish were sold, getting P875 in receipts.

Cash netflow

- The previous illustration of activities of the technology and the cash flows can be summarized by: (1) adding all money required to do the technology in a particular month to get the total monthly cash outflow; and (2) adding all the money received from the sales of the product of the technology in a particular month to get the total monthly cash inflow.
- Draw another calendar showing each month included in the previous calendar.
- Plot the total cash outflow by month on the lower part of the calendar.
- Plot the total cash inflow by month on the upper part of the calendar.

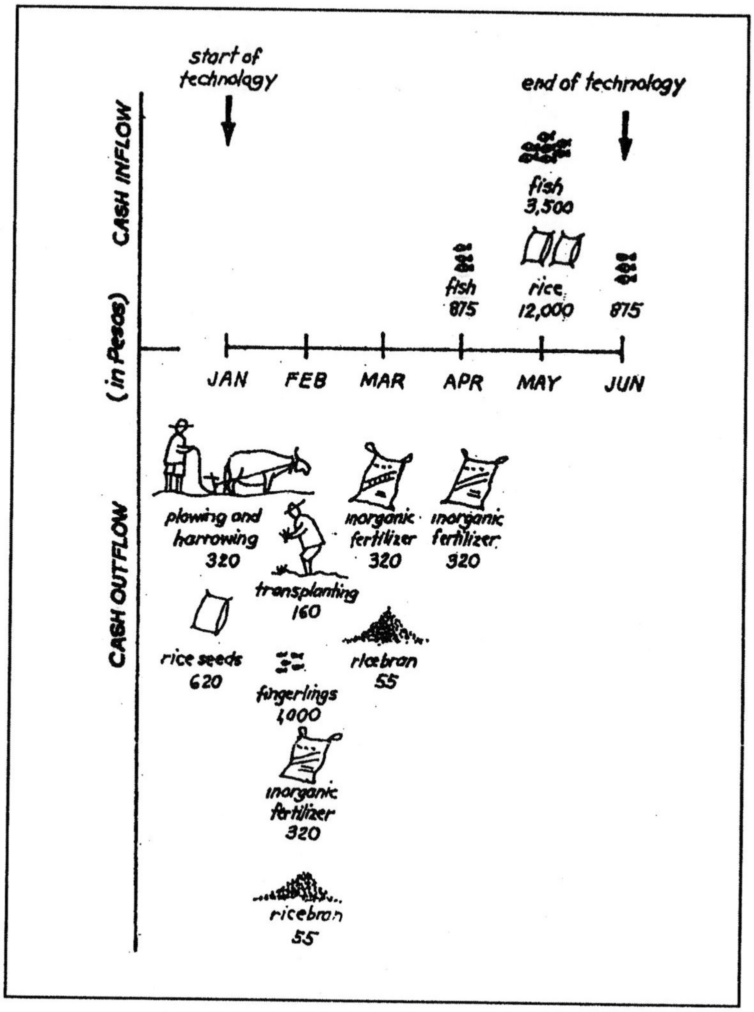

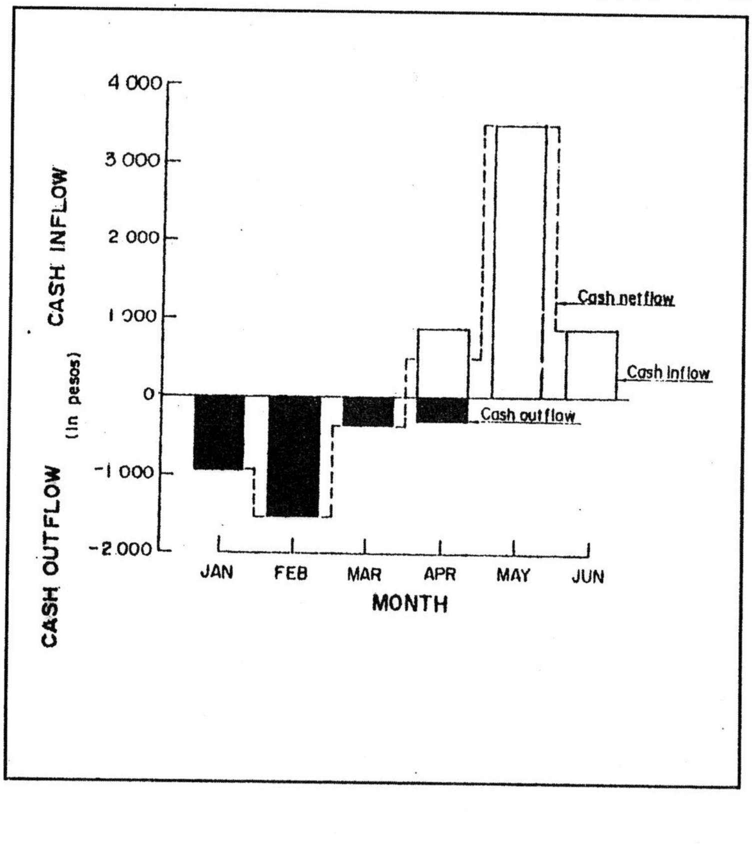

- The cash netflow is computed by subtracting the cash outflow from the cash inflow.
- A negative cash netflow, particularly the case in the first few months of the technology, means that the farmer spends money to buy and pay for things that are required by the technology. If he starts getting cash inflows, a negative cash netflow implies that more cash is required to pay for the technology than what is received from the sales of his products.
- A positive cash netflow implies that the farmer receives money from the sales of the products of the technology. When there are cash inflows and cash outflows in a particular month, a positive cash netflow means that the farmer receives more cash from the sales of his farm products which are able to pay for the farm expenses at that particular month.

Other economic considerations

- The farmer may have several alternatives in using his resources such as labour, land or cash capital as shown in the diagrams below.
- Before adopting a new technology (for example, rice-fish farming), the farmer would like to know whether using his resources for rice-fish farming would give him better income than investing these in other alternative income-generating activities.
- When the farmer has alternative uses for his resources, he should choose the activities that will generate more income from using his resources.

Opportunity costs

The opportunity cost of a resource (for example, labour, land or cash capital) is the value of the best alternative use of that particular resource. A new technology is worth adopting if the income earned from the use of the farmer's resources is greater than the opportunity cost (or what could have been earned) in other activities.

For example, a farmer's wife spends more time in the farm feeding the fish with rice bran and cleaning the dikes instead of cooking at home for the family. Children also help in the farm chores, thus spend less time studying school lessons.

Risks and market

- Is the produce from the technology meant for household and local consumptions or for export?

Heavy insect /disease damage to the rice crop will result in poor yields. Income from rice may not even be enough to recover farm expenses. However, as the fish are kept safe in the pond refuge, sales from them relieve this situation.

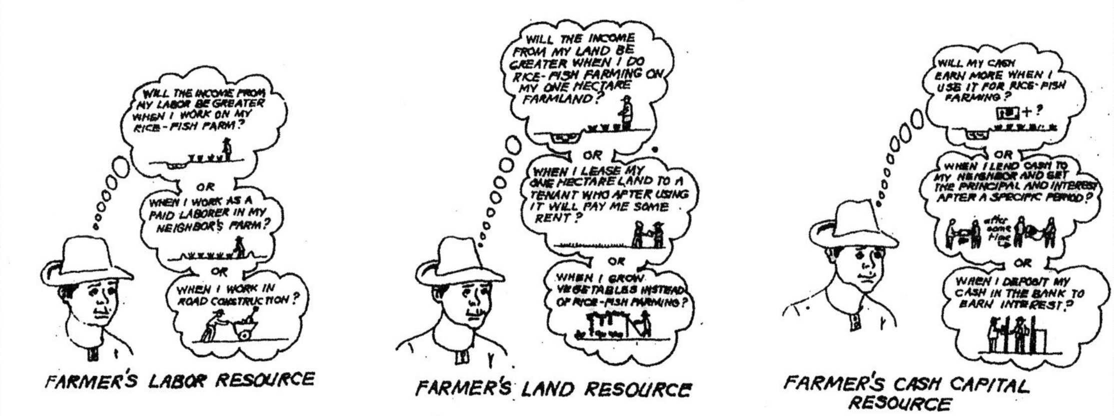

- How diversified will the farm operations become when the new component technology is adopted? Will it increase/reduce risks in crop failure?
- Will the products of new technology be subjected to high degree of price uncertainty because of unstable market? How sensitive is the net return to changes in input costs and output prices?

Equity/income distribution

Is it going to place significant demand for labour time from family members? Who will meet such labour demand? What is the opportunity cost of additional labour hours in terms of leisure, children's schooling, household work by female labour force, etc.?

Issues for further consideration

Aside from the "cash" orientation of the example presented here, other types of assets are used or gained through the adoption of fish culture. For farmers, of further importance are the resources saved by integration of fish with livestock, i.e. savings for feed, labour, etc., or gains to livestock or crops by introducing fish. For example, this can mean that the pond's capital cost is shared in terms of an irrigation supply for vegetables and drinking water for livestock.

In terms of livelihood options for small-scale households, how can they assess what alternative ways are there to improve family income and nutrition with minimum possible investment, least dependency on purchased inputs and lowest risk? Proposed farming options need to consider these fundamental issues.

Working with new entrants to integrated agriculture-aquaculture

Reg Noble and Clive Lightfoot

Developing integrated agriculture-aquaculture (IAA) systems for smallholder farmers requires their participation. This is crucial because farmers are the ultimate designers and managers of farming systems.

Often, smallholder farms are highly complex mixtures of crops, trees and livestock, which vary seasonally, using a range of resources and cultivated ecosystems. With such a diverse and difficult set of conditions, field extension-workers are often confused as to where and how to start.

One possibility is to utilize a very simple farmer-to-farmer technique that enables farmers to draw models of their farms with the help of other farmers and extensionists. The importance of this exercise is that farmers learn by doing.

The objective of drawings is to use this medium as a means for farmers to visualize their farm system so that they are better able to see new possibilities for integrating farm enterprises. These could be integrating new enterprises into the farm system or creating new linkages between existing ones.

Hopefully, there can be followup drawings with farmers to see how their farm systems evolve as they adopt new integrations.

Field exercise

The most appropriate setting for this exercise is in the farmer's own environment on the smallholding or in the village. Usually, it is better to start with groups rather than individual farmers.

Not only do groups allow more people to participate but also provide better dynamics than individual interaction when trying to make new entrants aware of different types of farm integration.

Group composition is also important. Mixed groups, which include women, men and children often work very well. However, the facilitator needs to ensure that individual interests do not dominate the gathering. In this context, it may be useful to have followup visits with single gender groups to see if viewpoints differ. You may choose to target groups of farmers who are likely to benefit from certain forms of integration. Rice farmers would be a suitable group for discussing rice-fish integration, for example:

- Cordially greet your farmer group and introduce yourself to everyone in the manner appropriate for the cultural setting.
- Explain that you have come to learn and understand how the farmers traditionally manage their farms.

- Suggest that they take you for a walk around the village or farm so that you can better understand their agricultural setting. Walking around and chatting in a relaxed atmosphere allow farmers time to relate their experiences. Thus, social distance and communication barriers are reduced. Do not take notes during this walk, just listen.

- Once the walk is over, continue the discussion. At an appropriate time, explain that with so much information, you find it difficult to visualize the whole farm system. Suggest to them that it would be easier for you to understand their farms if they could be represented in a drawing.

- If the farmers are agreeable to this idea, then explain carefully how to proceed with a drawing. Describe how actual plant or animal material can be placed on the ground to symbolize individual enterprises.

- Once farmers grasp this idea, then introduce the idea of linkages between enterprises with arrows. These arrows can be scratched out on the ground with a stick or marked with ash from fires. By doing this exercise for themselves, farmers learn more quickly the possibilities for integration on their farms.

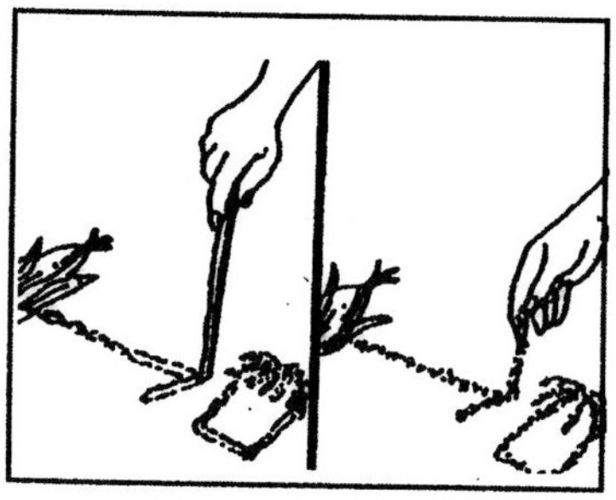

- Farmers should be allowed to interact among themselves so they can exchange ideas and produce a picture through joint effort. This group effort enables farmers to quickly learn from each other a range of ways of integrating farm enterprises.

If several farmers draw their farm systems together, drawing becomes a valuable tool for exchange of ideas between peers. Interchange of ideas facilitates generation of new ideas among farmers.

- The final drawing should show the full range of enterprises on the farm and linkages between them. This conveys farm integration more effectively than either written or spoken word. A picture of the farm system helps farmers appreciate their own farm as an integrated unit of interlinked enterprises.
- Finally, the farmers should be encouraged to consider how new linkages, inputs (on-farm and off-farm) and enterprises might be included on the drawing. Once a picture is drawn, it is easier for farmer/researcher/extensionist to see the possibility of making new links.

If a new enterprise is being introduced, it can be added to the drawing so the picture becomes the medium through which to discuss possible effects on farm operations. Drawings of individual farms enable the extensionist to see how integration varies from farm to farm.

- Drawing on a regular basis enables extensionists and farmers to follow through in a stepwise fashion the evolution of integration.

If farmers expand their drawing to include the whole village area, then common property resources can also be identified which have potential for linking to farm enterprises, such as aquaculture, livestock, etc.

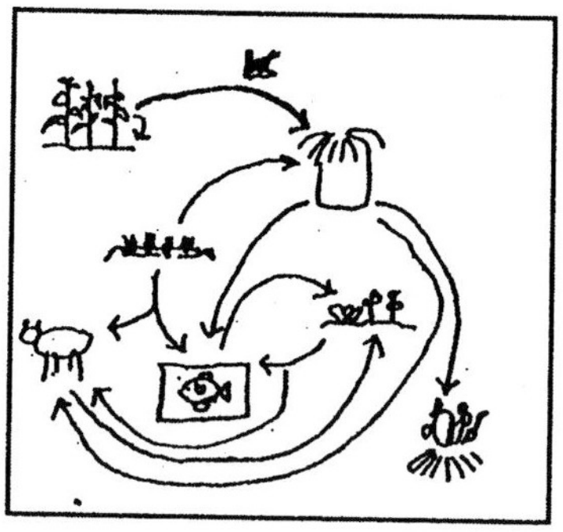

Summary

Incorporation of new enterprises, such as forestry and aquaculture, requires careful integration into traditional farming systems so that food security and income are not disrupted.

By drawing farm systems, farmers are better able to understand how new enterprises can be

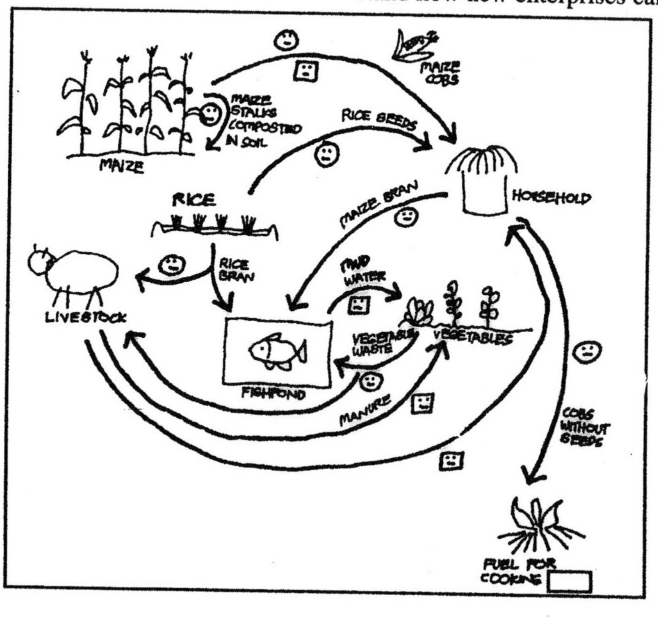

slotted in and enhance production of current enterprises with minimum disruption. Farmers can also evolve new integrations and management systems for themselves when they visualize their whole farm in a drawing.

Farm diagrams can also provide information on labour allocation with regard to gender. In the diagram above, simple symbols indicate whether men or women or both are moving resources around.

General principles for working with farmers

- Do not arrive on the farm at a bad time. One should check that farmers could receive you at the times you propose. This is particularly important when one wants to include mixed gender groups where women have different daily routines to men.
- Do not arrive on the farm with a large number of colleagues. This not only intimidates the farmers but also denies you the value of arranging your interview so that you empower women and other disadvantaged groups to speak out. More knowledge and experience are gained where a few interview many.
- Do not arrive on the farm in city clothes and giving orders. This only serves to increase the distance between you and the farmers. Your attire and attitude are powerful signals to rural folk; what they say is largely determined by how close you can get to them.
- Do not rush the interview. This usually results in you reconfirming what you already know because the unhurried exploration for new insights and cross-checking has not been possible. Relax, listen more than you talk and show respect of their knowledge by following up on leads offered by the farmers.
- Do not force your agenda. Our overriding concern to get the output needed and conclude the interview quickly reduces the quality of our data and our relationship with the household. Rather, we should let the information emerge naturally. Forcing farmers to draw diagrams not only results in you drawing the diagram for them, but also in the farmers finding little value in them. This makes it difficult for you to return. If farmers learn from the interview, they will invite you back.
- Do not continue on with a bad interview. When, for any number of reasons, you find yourself interviewing farmers who are distracted by other matters, as happens to us all, recognize the fact and tactfully withdraw. It is better for you and others that follow you to have good relationships with the community rather than good data on the community.
- Do not appear with paper and pens and instruct them to draw a picture of their farms. This will not work. The picture must emerge naturally as a way for the farmers to express all that is happening on their farms.
- Explain to the farmers that they are the teachers in this exercise and you, the extension worker, are the pupil. This shows respect for the farmers' knowledge and provides a more equitable working relationship between the visitor and the farmer.
- It is important to encourage farmers to use their own methods and materials to represent farm enterprises. The visitor should avoid doing any drawing; otherwise, the farmers might be intimidated and withdrawn.

Issues for further consideration

Experience has shown that the methods described here are useful in new situations for an "outsider" to understand any farmers' system, not just new entrants. The approach has also worked well for farmers to learn how they can further improve an existing fish production system they have managed for some time.

On the other hand, the approach is very time-consuming if used on a wider scale, e.g. in extension efforts. In this respect, there are roles and experiences with mass media and mass organizations for large-scale extension activities.

In this process, there may be potentially different roles for fisheries-trained persons and nonspecialists. Having multidisciplinary teams in the exercise has proven valuable.

Users should be prepared for communication problems between outsider and farmer, and how to deal with the situation. The role for translation needs to be considered, not just for foreign nationals but also for nonlocal language speaking nationals.

In cultures where gender, caste, class and ethnicity prevent communities from meeting together at one level, alternatives for application need to be designed. The comment that "mixed groups that include men, women and children" often work well is not universally applicable.

The application of this approach to community-managed fishponds or collective fish culture needs to be assessed and compared with other methods.

Integrated agriculture-aquaculture and the environment

Roger Pullin

General considerations

Food production invariably has environmental effects: occupation and fragmentation of former natural habitats; reduction of the abundance and diversity of wildlife; and changes in soil, water and landscape quality. Most integrated agriculture-aquaculture (IAA) systems use low levels of inputs and fall within the type of aquaculture called semi-intensive. This means less reliance on heavy feed and fertilized inputs, lower densities of farmed organisms and, therefore, less chances of causing serious pollution and disease risks than more intensive, feedlot-type systems. This is important as it is the high output of the foodstuffs necessary for intensive feedlot systems that create environmental pollution. Semi-intensive systems in synergy with agriculture (crop-livestock-fish integrated farming) capitalize on *in situ,* vitamin and protein natural aquatic feeds, which obviate the need for expensive feed components.

Semi-intensive freshwater ponds usually have few environmental effects other than their occupation of former natural habitats. In the tropics, where there is fast turnover of organic waste loading, their effluents and excavated muds usually enhance the productivity of adjacent waters and lands and avoid overenrichment.

Special care is needed, however, where pond and dike construction may disturb acid sulphate subsoils and where water table changes may uplift subsurface salts. Moreover, saltwater intrusion from coastal ponds may poison soils and freshwater aquifers. The use of chemicals in semi-intensive aquaculture is usually limited, but farmers should always take great care when using antibiotics, hormones and other drugs and should follow the instructions very closely. Seek professional advice from a veterinarian or fish culture specialist and be aware that many drugs are persistent in the environment.

Choice of fish species

The aquatic medium is shared by many users and supports a diverse fauna and flora. As aquaculturists develop better domesticated breeds, international demand for these will increase. This means increased transfers of exotic breeds, as has been of immense benefit for crop and livestock farming. However, cultured aquatic organisms often escape and form feral populations which may: (1) displace or interbreed with wild stocks, thereby threatening natural genetic resources; (2) disrupt natural habitats by causing proliferation or clearance of vegetation or increasing turbidity (benthic foraging); and (3) introduce aquatic pathogens, predators and pests inadvertently.

Development agencies and farmers must weigh the benefits of using exotic breeds against possible environmental consequences. Development projects and farmers often try exotic breeds without thorough appraisal of the possible consequences. Such irresponsible experiments may have far-reaching consequences; loss or damage to habitats and genetic resources of wide importance. This damage may last forever. Codes of practice to avoid this have recently been developed, but aquaculture development still lags behind agriculture in recognition of the risks of transfers and international application of these safeguards.

The only general guidelines here are: (1) use native species and breeds developed by local or national programs wherever possible; and (2) if the introduction of other species or breeds needs to be considered, seek professional advice on how to assess the possible consequences and comply with the laws and Codes of Practice that have been developed for the good of all present and future farmers.

Public health

IAA generally has no special health risks significantly greater than agriculture, but freshwater ponds may assist the spread of waterborne diseases. They can harbour the intermediate hosts of parasitic worms, such as bilharzia, and can be breeding sites for mosquitoes. Such problems are minimized by maintaining weed-free, well-stocked ponds. In fact, many species of fish eat and control mosquito larva but snail control by fish is not usually possible.

Fish farm workers who enter ponds may risk bilharzia infection in infected areas and other waterborne microbial diseases (viral, leptospiral, bacterial and fungal).

On the positive side, many of the pathogens and parasites that contaminate fish produced from livestock excreta-fed ponds are eliminated by a well-fertilized pond environment, as in sewage oxidation ponds. Problems of pesticide accumulation in ricefield fish are diminishing because of the increased use of integrated pest management programs employing natural substances and predators.

The risk of accumulation of heavy metals from livestock feeds in manured pond sediments and fish is slight and applies more to intensive systems. The same probably applies to pathways for aflatoxins (poisons that develop from fungi in badly stored feeds) but this has been little studied.

Sewage-fish culture is controversial because of assumed health risks to farm workers and fish consumers. However, these may be slight compared to the nutritional benefits provided that postharvest handling of the fish is hygienic (with particular attention to not rupturing the gut and allowing its contents to make contact with fish flesh). Such produce must also be well-cooked.

There are no general guidelines on how to minimize these risks other than to be aware of which waterborne diseases are present in any given locality and to assess whether the establishment and operation of ponds significantly adds to the risks of contraction by farm workers, fish handlers and consumers.

Seek professional advice from public health workers.

Issues for further consideration

Integration of farming activities through recycling is a traditional practice in many societies in Asia, where these have evolved beyond slash-and-burn cultivation. On the other hand, the integration of aquaculture into these traditional farms was and still is not widely practiced, counter to common perception. It originated in China and was spread throughout many parts of Asia by Chinese migrants, but often remained a practice common only to these groups.

The multiple-use aspect of ponds and their benefits to other enterprises of the farm, notably vegetables and livestock, is characteristic. Here, the impacts of semi-intensive ponds on the environment, through conservation of surrounding habitats and species, may be positive. The conversion of low-lying areas to perennial ponds can benefit surrounding natural habitats and organisms by increasing the availability of water.

Snail control may be one of the benefits of integrating fish culture closely within the farming system. Ducks allowed to forage in paddy fields and fishponds can control snails, as well as other pests. If done correctly, catfish and common carp stocked in ponds and ricefields will keep golden apple snail damage below economically damaging thresholds.

Taking a different perspective, the impacts of the environment on the potential and practice of IAA should be considered. For example, the potential for IAA may be high in peri-urban areas, but with high prevalence of pollution from industry and households, increasing competition for water and land at increasing prices, high rates of city expansions covering agricultural land, the probability for sustained success of such activities is low.

INTEGRATED FARMING SYSTEMS

Integrated grass-fish farming systems in China

Huazhu Yang, Yingxue Fang and Zhonglin Chen

Introduction

Integrated fish farming systems refer to the production, integrated management and comprehensive use of aquaculture, agriculture and livestock, with an emphasis on aquaculture. China has a long and rich history of integrated fish farming. Written records from the first and second centuries B.C. documented the integration of aquatic plant cultivation and fish farming.

From the ninth century, records showed fish farming in the paddy field. From the fourteenth to sixteenth centuries, there were records of rotation of fish and grass culture; and by the 1620s, the mulberry-dike fishpond, the integration of fish and livestock farming and complex systems of multiple enterprises integrated with fish farming were developed.

Integrated fish systems, using grass and aquatic plants as fish feeds, are commonly found in many parts of China. These systems are particularly predominant in the irrigated lowland areas of the Changjiang, Pearl and Yangtze River basins. Many of these farms are large, communal ones with co-operative or collective farming, which are commonly found throughout China. Farmers in these areas grow graminea species in various areas of their farms, including fields, small plots of unused land, pond dikes and drained ponds. The grass is then fed directly to the fish as a supplemental feed. In southern China, farmers also use available water resources, such as rivers, lakes, ditches and pools, to cultivate aquatic plants for their use as fish feeds.

Three integrated systems from China, involving grass and/or water hyacinth are presented here: grass-fish, water hyacinth-fish and pig-grass-fish.

Integrated fish farming systems

Grass-fish

Grass species, which can easily be produced on the farm, can serve as low-cost supplemental feeds for fish. Commonly cultured fish species, which can feed directly or indirectly on grass, include grass, silver, bighead and common carps. As seen in Figure 1, grass can be grown along pond boundaries and fed directly to fish. Grass species commonly used include rye, Sudan and napier grasses (see Table 1). Figure 2 outlines a seasonal calendar for grass production within a grass-fish system. The information presented here excludes data for labour and opportunity costs.

Pond sizes range from about 0.5-1 ha in size with water depths of 2-2.5 m. Net fish yields of up to

Figure 1. Bioresource flows for grass-fish system

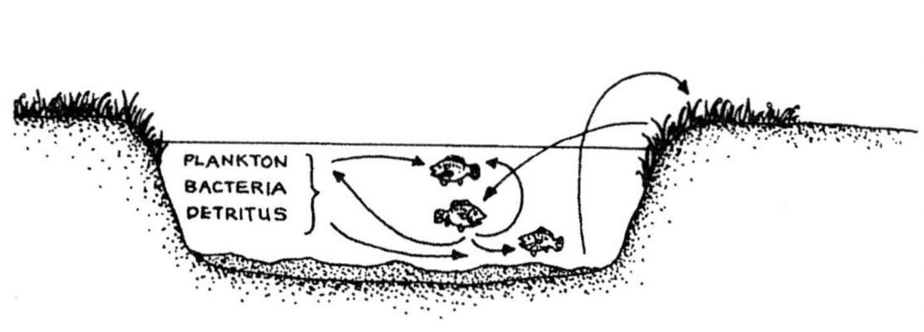

Figure 2. Seasonal calendar for fish and grass/water hyacinth production

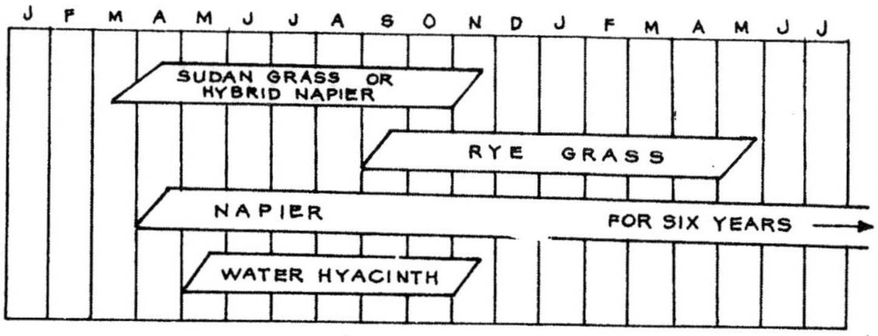

Table 1. Summary of important aquatic and terrestrial species used for grass-fish integration. Parts used are leaf and stem. Conversion factor is amount of fresh grass required to produce 1 kg of fish.

Species	Yield (average fresh weight in t/ha)	Conversion factor
Rye grass *(Lolium multiflorum)*	75-150	17-23
Sudan grass *(Sorghum vulgare* var. *Sudanese)*	150-225	19-28
Elephant grass napier *(Pennisetum purpureum)*	225-450	30-40
Hybrid napier grass *(Pennisetum americanum)*	225-300	25-30
Water hyacinth *(Eichornia crassipes)*	150-300	45-50

6 t/ha have been recorded without supplemental feeding or the use of additional green or animal manures. An area roughly one-half the size of the pond is needed to produce sufficient grass for supplemental feeding. Figure 3 shows that on-farm produced rye grass and Sudan grass can be sufficient to meet fish production feed requirements. Rye grass and Sudan grass can yield up to 112 t/ha/season (fresh weight) and hybrid napier can yield up to 300 t/ha/season (fresh weight).

The use of cereal grains as feed supplements in fish production can be costly; using grass species can be much more economical. Production costs related to supplemental feeds are 50 percent lower (per kg of fish produced) for grass-fed vs. cereal grain (barley)-fed fish.

Figure 3. Monthly fish-feed requirement and average daily production of rye and Sudan grasses

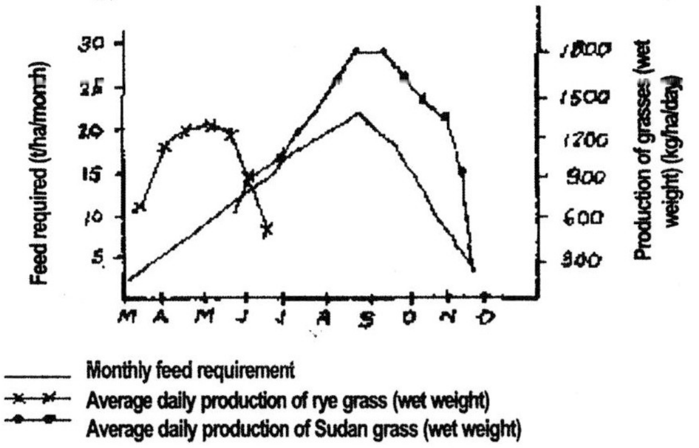

Grass-fed systems work well in China because: (a) competition for grass is limited as grazing animals are less important; (b) large grass carp seed are available; (c) grass carp are relatively valuable; and (d) other fish species are available to utilize grass carp wastes in polyculture.

Table 2. Stocking/harvest design for grass-fish system

Species	Stocking			Harvesting			
	Individual weight (kg/fish)	Density (fish/ha)	Total weight (kg/ha)	Survival (%)	Individual weight (kg/fish)	Gross yield (kg/ha)	Net yield (kg/ha)
Grass carp	0.50	2 000	1 000	95	2.5	4 750	3 750
Silver carp	0.05	3 200	160	95	0.5	1 520	1 360
Bighead carp	0.05	800	40	95	0.5	380	340
Common carp	0.05	2 400	120	85	0.5	1 020	900
Total			1 320			7 670	6 350

Water hyacinth-fish

A variety of aquatic plants can be used as supplemental feeds in fish production, among these is the water hyacinth. An area approximately one-half the size of the fishpond is needed to produce enough water hyacinth for supplemental feeding. Water hyacinth can produce up to 300 t/ha/year (fresh weight). Net fish yields can also reach 6 t/ha/year without supplemental feeding or use of additional manures. Pond sizes and stocking rates are the same as in the grass-fish system. Fish input costs using water hyacinth comprise less than 15 percent when compared to cereal grain (barley)-fed fish. Resource flows are outlined in Figure 4. Note that in many countries, water hyacinth is banned and has caused serious problems in lakes, rivers and estuaries.

Pig-grass-fish

Pig-grass-fish integration is widely practiced and has good economic returns depending on labour costs. Large-scale pig farms produce large amounts of excreta which, for the purpose of reuse and treatment, are used as fertilizer for high-yielding fodder grasses, which in turn are used as the main feed for herbivorous fish. Pig excreta are only partly applied directly to the fishponds. The excreta of herbivorous fish fertilize the pond water to support the growth of fish. The pond humus can then be used as manure for plant cultivation. Thus, the productivity of both fodder grasses and phytoplankton can be utilized.

Pig-fish and grass-fish components can be integrated to optimize the resource flows for increased productivity (Figure 5). About 45-60 pigs can support 1 ha of grass production per year (225-300 t rye grass and Sudan grass) for a 2-ha pond (6 t/ha/year of fish yield) (Figure 6).

Figure 4. Bioresource flows for aquatic plant-fish system

Figure 5. Nutrient pathways in pig-grass-fish system

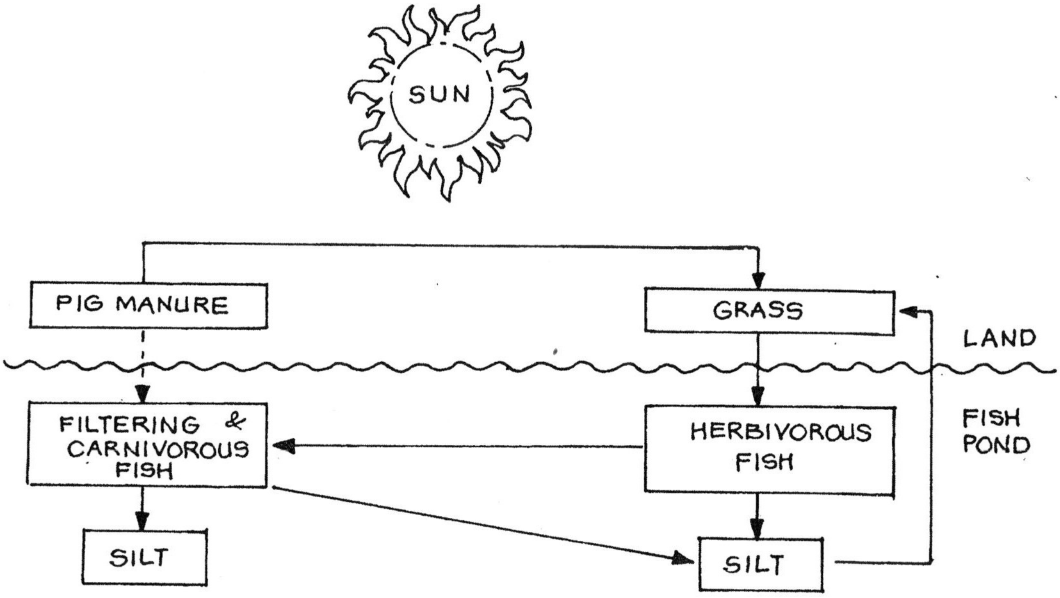

Figure 6. Fertilization calendar for grass production

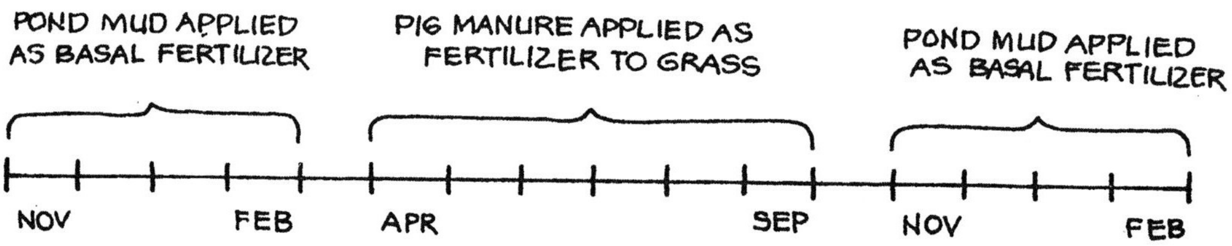

Issues for further consideration

The example shown here has been applied on large-scale state farms, where component enterprises were designed and managed to optimize production, and the required labour force could be appropriated as needed. The example is also widely used in small-scale, family-operated fish farms.

With the recent trends of increasing affluence and subsequent increase in cost of labour, change in market demand, changing needs of the farm households and availability of other livelihood options, there is significant impact on such practices. Many such farms in some areas of China have now switched to feed-based semi-intensive fish culture practices.

Initially the system was based on grass collection from various areas, but the technology was evolved and improved by intensification of grass production (both quality and quantity) and including better balanced fish polycultures, which may be appropriate for other recommendation domains given a similar context.

The relevance of the grass-based system for smallholders may be limited due to considerable requirements for space (adequate land area for growing large amounts of grass to adequately feed the fish) and labour (cutting grass for the fishpond, removing mud and broadcasting onto grass). The opportunity costs of the land for growing pasture grass rather than other crops must be evaluated according to the local situation.

The resource costs of growing the necessary high-quality pasture grasses (which is an important issue in this technology) need to be considered. All grasses need considerable and frequent fertilizer application to yield well and produce grass of acceptable nutritional quality for grass carp. Farmers will need information on how to optimize grass production and quality.

Adopters will need to consider how the labour demands for collecting and growing grass can be managed within the household. In northern Vietnam, women do most of the grass cutting and spend up to 2-3 hours per day on this activity.

In locations in which macrophagous (i.e. grass-eating) fish are nonexistent, alternative management methods can be recommended. Of consideration is also, how much of the grass works as direct feed for the grass carp, and how much as green manure to the pond ecosystem.

Research has shown that water hyacinth has poor palatability for fish compared to other aquatic plants. Usually water hyacinth, which is a banned pest and environmental hazard in many countries, grows in unutilized ponds, ditches and common property waterbodies. The case presented here indicates options to utilize this resource. The weed can also be used to remove nutrients, e.g. for wastewater treatment, although then accumulation of pollutants may become an issue.

Given larger amounts of available pig manure (e.g. from pig farms), and macrophagous fish in a polyculture, the recycling of pig manure onto grassfields, and the subsequent feeding of grass to fish may be an option. Otherwise it may seem more efficient to add the pig manure directly to the fishpond.

Chinese embankment fish culture

Kuanhong Min and Baotong Hu

Embankment fish culture, along with bamboo and mulberry culture, has been practiced in the Yangtze River delta and Pearl River delta areas of central and south China for centuries. Originally, the delta was just a waterlogged area. Farmers dug and moved soil, piling it into huge rectangular or round shapes and utilized these raised embankments for planting crops. The excavated areas became deeper, making them ideal for fish culture. Where embankments are wide enough, mulberry, bamboo, etc., can be grown (see Figures 1 and 2). The mud is scraped from the bottom of the pond and applied as fertilizer to the embankment 2-5 times annually at a rate of 750-1 125 kg/ha/year.

Figure 1. A view of embankment fish culture system

Figure 2. Farm transect of embankment model

LANDTYPE	POND	EMBANK-MENT	POND	EMBANK-MENT	POND	EMBANK-MENT
CROPS		MULBERRY BAMBOO		MULBERRY BAMBOO		MULBERRY BAMBOO
ANIMALS	FISH		FISH		FISH	
SOIL		CLAY LOAM		CLAY LOAM		CLAY LOAM

Mulberry plot-fishpond

In this system, which is more common in form of large-scale farming systems, the mulberry leaves are used as feed for silkworms. The sericulture provides a large variety of feeds and fertilizers for fish farming. On the other hand, the technology requires high labour input.

It has been determined that 36 700 kg/ha of mulberry leaves can be produced which can yield 2 700 kg of cocoons and 18 400-18 750 kg of silkworm excreta and silkworm sloughs (molted skins). The silkworm excreta can both serve as feed and fertilizer for fish. The cocoons contain 80 percent pupae by weight. The feed conversion ratio of pupae to fish is 2:1 such that 2 kg of pupae can produce 1 kg of fish. All the feeds and manure from silkworm farming can support a good fish yield; see Figure 3 for the cycling process of the silkworm wastes. The suggested fish stocking in mulberry plot-fishpond is listed in Table 1.

Bamboo plot-fishpond

The produce from bamboo farming is mainly bamboo shoots. Zhangchai Township, Fusan, Guangdong province, has long been processing canned bamboo shoots. It is estimated that 25-30

Figure 3. The cycling and conversion process of silkworm wastes

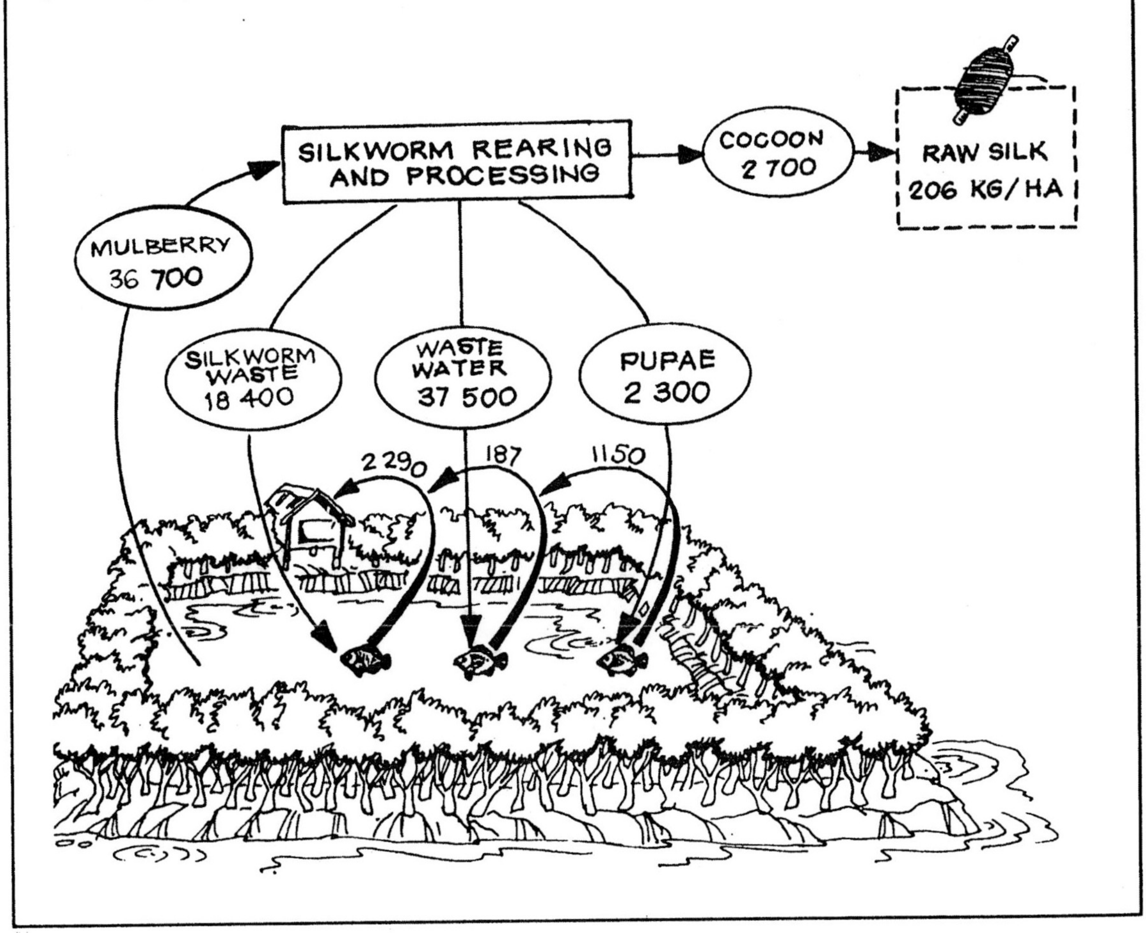

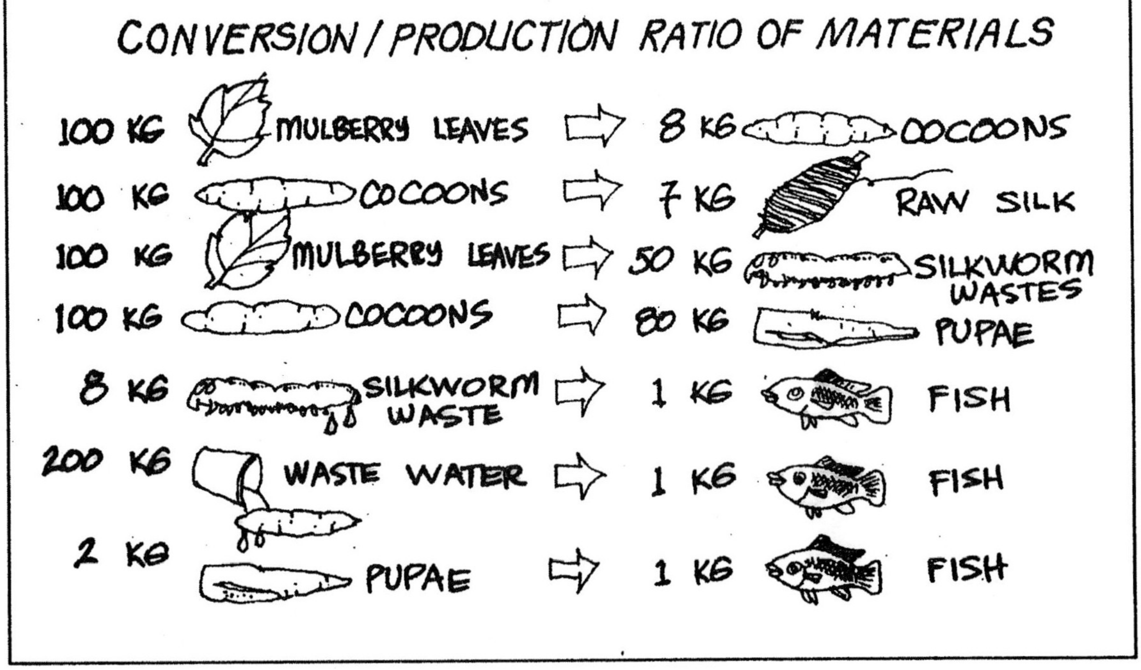

Table 1. Stocking for mulberry plot-fishpond

Carp species	Stocking			Survival (%)	Harvest		Body weight increment (factor)	Net yield (kg)
	Fish weight (g)	Total weight (kg)	Total stocking (pcs)		Fish weight (kg)	Number of fish		
Silver	50	97.50	1 950	90	0.75	1 755	12.50	1 218.8
Bighead	50	22.50	450	90	0.75	405	12.50	281.2
Grass	500	225.00	450	90	1.75	405	2.15	483.8
Common	25	18.75	750	85	0.60	637	19.38	363.4
Crucian	10	19.50	1 950	95	0.20	1 852	18.00	351.0
Total		383.25						2 698.2

percent of the wastes and by-products could be used for fish farming. Wastes and by-products from a 1 ha harvest of bamboo shoots can produce about 500 kg of fish.

A modest estimate from the farmers of Zhangchai Township shows that bamboo production per hectare ranges from 22.5 to 26.3 t/year. But when shoot production is over, the farmers harvest the old bamboo poles, totaling 52.5 to 67.5 t/ha/year. These can be used as firewood, construction materials for livestock pens or support materials for climbing plants (see Figure 4).

The mud from the bottom of the pond provides a very large amount of compound fertilizer for the bamboo plot. In shoot production, 6 000 kg/ha of pond mud containing 168 kg N, 109 kg P and 150 kg K are needed, but one-fourth of the supplied amount of pond mud is more than sufficient to supply the needed nutrients. Therefore, the total amount of nutrients supplied by the mud cannot be fully absorbed by the plants. The farmers in Zhangchai Township realize that the shoot production is 20-30 percent higher in the pond plot than in hilly areas, probably because of good ventilation in between plants and adequate water and fertilizer supply. Mud application moreover impedes the growth of wild plants and improves the soil quality. Bamboo plot-fishpond stocking rates are illustrated in Figure 5. Figure 6 shows the farming calendar in fish-sericulture-bamboo production.

Figure 4. Material flow in bamboo plot-fishponds

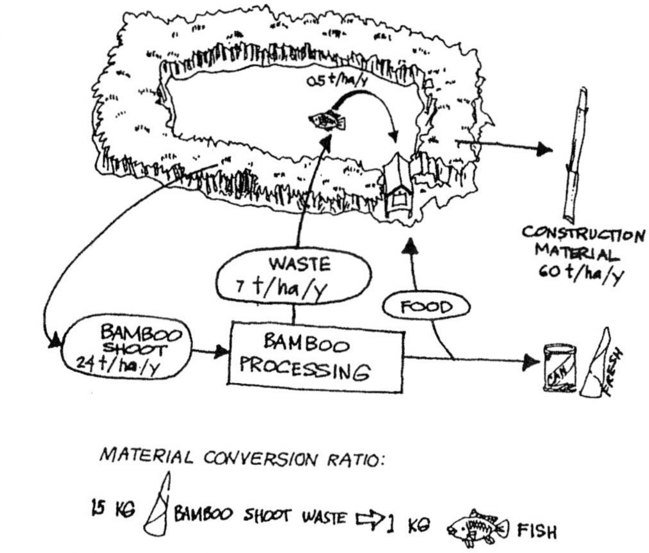

Figure 5. Suggested stocking rates for bamboo plot-fishponds

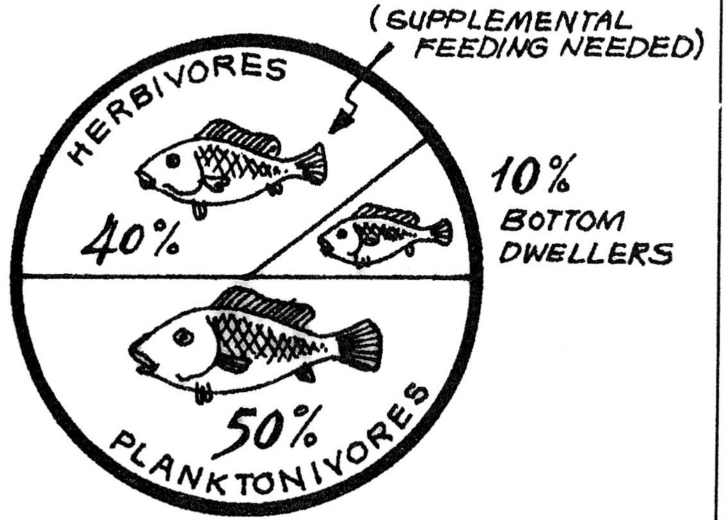

Figure 6. Farming calendar in fish-sericulture-bamboo production

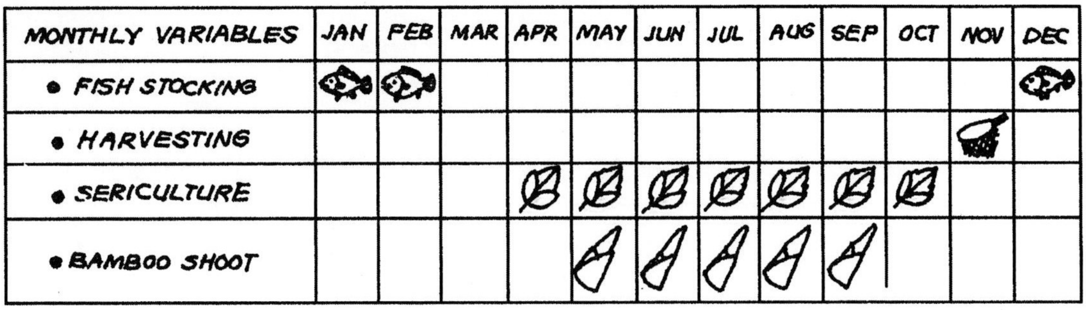

Issues for further consideration

The mulberry-fish system has declined with industrialization in many areas of southern China because of opportunity costs of land and labour. The bamboo-fish system is dependent upon a processing industry in the vicinity. In most cases, the bamboo shoots are not the major crop.

Both mulberry and bamboo systems are unusual, compared to many potential embankment crops, in that they are perennial crops. The considerable amounts of mud that need to be removed from the pond for dike crop fertilization require considerable amounts of labour.

In the last ten years the practice is losing its popularity due to economic changes manifested by increased cost of labour, shifts in market demand, changes in food preferences and availability of other livelihood opportunities.

The mulberry-fish system requires high labour inputs. A silk processing factory should be in the vicinity.

For bamboo shoot production, there should be adequate market demand, processing facilities in the area, high rainfall and a year-round humid climate.

Required nutrient inputs need to be verified depending on soil quality and plant requirements. It is unclear if additional fertilizer amounts are given to the dike crops, aside from the pond mud.

The VAC system in Northern Viet Nam

Le Thanh Luu

The Viet Namese saying *Nhat canh tri, canh vien* states that the first profitable activity is aquaculture and the second is agriculture, horticulture or gardening. Integrated farming is a traditional approach to family food production in the poor, rural regions of Viet Nam. The integration of the homelot, garden, livestock and fishpond is called the VAC system (VAC in Viet Namese is *vuon, ao, chuong* which means garden/pond/livestock pen).

The widespread promotion of the VAC system, referred to as the VAC movement, began in the early 1980s. The importance of small-scale integration was emphasized by the late Pres. Ho Chi Minh in the late 1960s. The objective of the movement was to increase and stabilize the nutritional standard of the rural poor. Because of adoption of the VAC system, the dietary standard of the rural poor significantly improved, particularly in the isolated villages in the high mountainous regions.

This farming system is family-managed, with practically all labour coming from the household. VAC farms can be found in various agroecological conditions, including irrigated lowlands, rainfed uplands and peri-urban areas (Figure 1). Ponds are usually constructed primarily for raising land for the home and garden. Traditionally, the water collected in the de facto pond was used for domestic purposes and to produce aquatic weeds for feeding to pigs. Most pig and other manures are used on field crops, especially rice, although as fish production grows in importance, more is diverted for use there.

It is estimated that 85-90 percent of the rural families maintain a garden and livestock pen, with 30-35 percent of these having fishponds. In many villages, 50-80 percent of families have the full VAC system. Figures show that 30-60 percent of income of most village families may come from the system; in many cases, it may be 100 percent.

Upland VAC system

The upland VAC system (Figure 2) is usually found in mountainous regions, such as Hoa binh, Son la, Ha giang, Tuyen quang, Thais nguyen and other provinces.

Figure 1. Seasonal calendar of agriculture-aquaculture activities in the uplands and lowlands

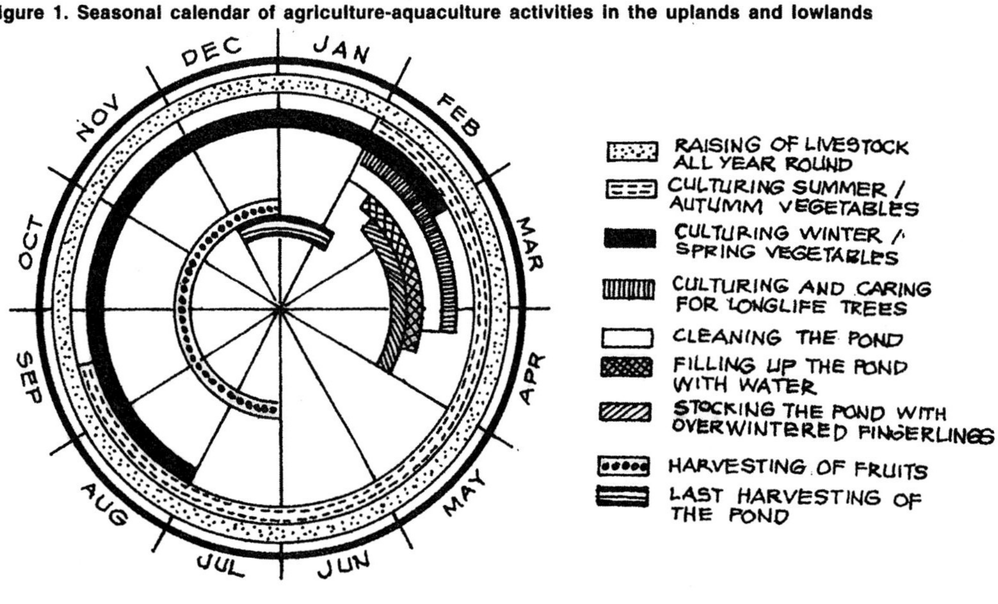

Figure 2. Upland integrated farming system

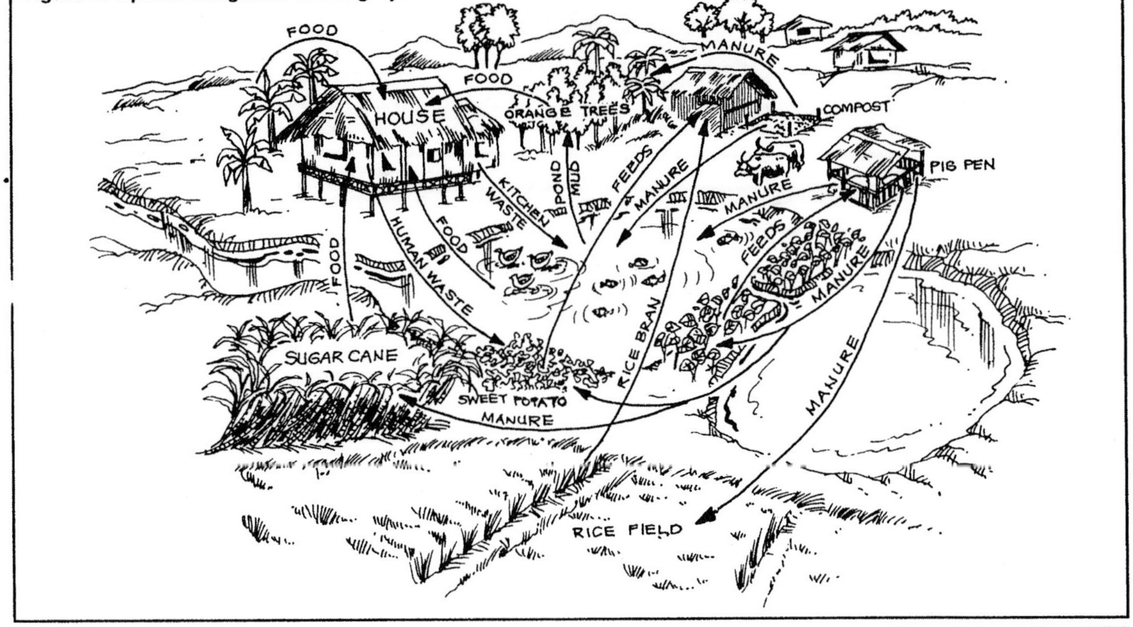

Figure 3. Lowland integrated farming system

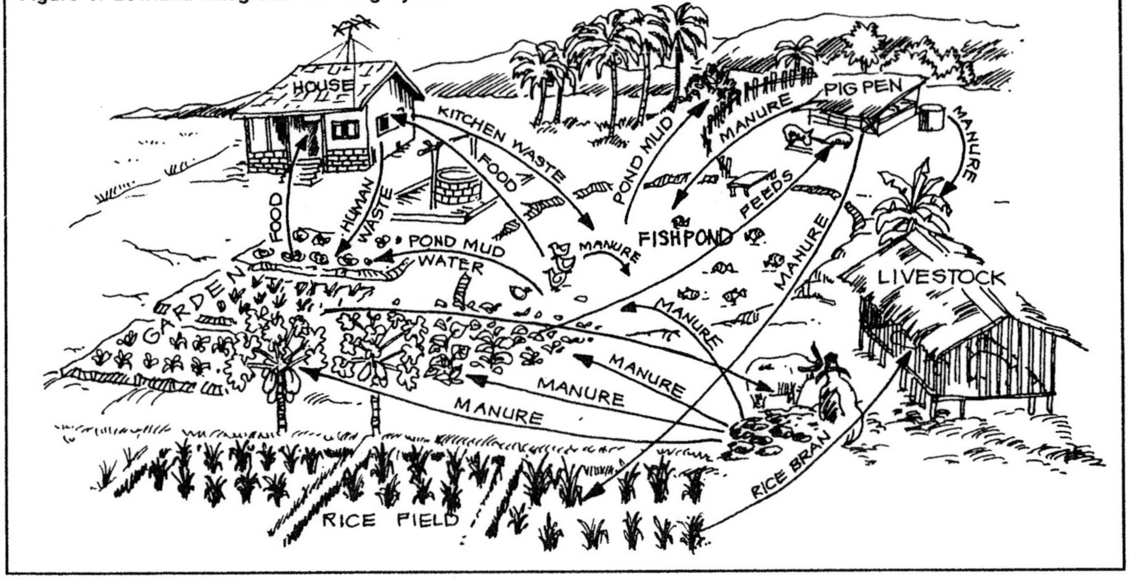

The pond is constructed close to the house so that the domestic and kitchen wastes are drained into the fishpond. The livestock pens and garden are also situated near the pond. The 1 000-5 000 m^2 garden includes a variety of vegetables (i.e. green onion, sweet potato, watercress, etc.) and fruits (i.e. banana, orange, peach, apricot, etc.) and other crops, including sugarcane, tea and cassava. This provides a mix of perennial and annual crops.

A portion of the livestock manure is used for manuring the trees and vegetables. Trees are manured once or twice a year; vegetables are manured according to their needs. Pond silt is removed every 3-4 years and used as fertilizer.

Most families keep various animals on the farm, including one or more water buffaloes and cattle, one or more pigs, and several ducks and chickens. The large ruminant animals are allowed to graze or are fed farm by-products. The swine and poultry are usually fed with kitchen wastes, as well as other farm by-products, such as cassava, rice bran, sweet potato, banana trunks and water hyacinth.

The fishpond is usually allocated a more central part of the farm for better management. Pond area ranges from 100-1 500 m^2, with a pond depth of about 1 m. Ponds are often drained after the final harvest, usually in February. The pond bottom is kept dry for 1-3 weeks; after which it is cleaned, limed, manured and then

Basic features of the integrated VAC system in Northern Viet Nam

Basic features	Upland	Lowland
Garden		
1. Area	1 000-15 000 m²	200-300 m²
2. Cultivation practices	Perennial + seasonal planting in March	Perennial + seasonal
- Fruit trees	Seasonal cultivation	
- Vegetables	Seasonal cultivation	Seasonal cultivation
- Rice		Seasonal cultivation
3. Manuring		
- Fruit trees	Pond mud	Pond mud
- Vegetables	Livestock manure + human wastes	Livestock manure + human wastes
- Rice	Pig manure	Pig manure
4. Number and type of livestock		
- Buffalo	1-3	1-2
- Cow	1-6	1-4
- Pig	1-3	1-2
- Chicken and duck	Several	Several
5. Feed resources		
- Buffalo and cow	Grass, rice straw, rice bran, sweet potato	Grass, rice straw, rice bran, sweet potato
- Pig, chicken, duck	Rice bran, cassava, kitchen wastes, sweet potato, banana, water hyacinth	Rice bran, cassava
Fishpond		
1. Area	100-1 500 m²	50-400 m²
2. Average depth	1 m	1.0-1.2 m
3. Stocking rate	Silver carp: 20-25% Grass carp: 5-10% Common carp: 5-10% Rohu: 20-30% Mrigal: 20-30%	Silver carp: 25-35% Grass carp: 2- 5% Hybrid common carp: 10-15% Rohu: 20-30% Mrigal: 15-25%
4. Stocking density	0.5-2.0 fingerlings/m² (3-6 cm)	1.0-2.0 fingerlings/m² (5-6 cm)
5. Manuring	Kitchen wastes, livestock manure (0.05 kg/m², twice per month), green manure	Kitchen wastes, livestock manure
6. Harvesting	Continuous harvest, after 3 months of culture	Continuous harvest
7. Estimated production	1.0-1.2 t/ha/year	1.5-2.0 t/ha/year

filled up with water for restocking. Domestic washings and kitchen wastes are channeled into the pond daily. Animal manure is also applied twice a month at the rate of 0.05-0.15 kg/m². Three months after stocking, farmers begin to harvest on a weekly basis using small nets and continuously restock and harvest the pond.

Lowland VAC system

The lowland VAC system (Figure 3) is usually found in Ha noi, Hai duong, Hung yen, Ha nam, Nam dinh, Hai phong and other provices.

In the lowland areas of North Viet Nam, the integration of garden, livestock and fish culture is also common. Usually, houses are constructed close to the pond. In sandy regions, the houses are often built at some distance from the pond for hygienic reasons.

The garden is usually small, 400-500 m². Fruit crops commonly grown include banana, orange, papaya, peach, litchi, longan and apple. In many suburban family gardens, ornamental trees and flowers are planted as a main income source. Vegetables grown include green onion, sweet potato, cress, tomato, cabbage and water spinach. Both perennial and annual crops are planted to provide year-round food to the house and products for the market.

Pond mud is annually removed and used to manure fruit trees; livestock manure is used to fertilize vegetables. Pond water is used for irrigating the garden, especially vegetables.

Most families keep various animals on the farm, including one or more water buffaloes and cattle, one or more pigs, and several ducks and chickens. The large ruminant animals are allowed to graze or are fed farm by-products. The livestock pens for pigs, buffaloes and cows are constructed at the corner of the garden close to the pond. The swine and poultry are usually fed kitchen wastes, as well as other farm products and by-products, such as cassava, rice bran, sweet potato, banana trunks and water hyacinth.

Most families have ponds of 50-400 m², with different shapes and an average depth of 1.0-1.2 m. Ponds are drained after the final harvest (usually in January/ February). The pond is then kept dry for a few days, limed, manured and refilled with rainwater or irrigation water (early rains may start at the end of March.) Domestic washings and kitchen wastes may be channeled into the pond with a small part of the manure coming from the livestock used to manure the pond (according to the farmer's experience). Leaves of legumes, such as peanuts, green beans, etc., are also used for manuring the ponds.

Issues for further consideration

In recent years significant improvements have been made to the VAC system with significant increases in fish yield. In upland areas, farmers have also been able to manage stream flows and make them pass through series of VAC ponds where significantly higher fish yields are achieved. The improved practice has potential for application in several other countries in the region, where farmers have relatively small farm holdings.

Human waste is a critical agricultural input in the VAC systems, aside from on-farm produced livestock feed, particularly rice bran and sweet potato. The complementary linkage of all farm components can best be studied with a cropping calendar that includes fish, rice, livestock and other crops.

The lowland systems are significantly different from the upland systems in terms of resource availability and use. Three systems are distinguished: (a) suburban, (b) intensive (i.e. rice production) and (c) lowland (i.e. floodprone environments).

The VAC systems are traditional and farmers and researchers have developed more intensive approaches over the last decade, since economic changes have taken place. They are less specific in their application requirements than Chinese systems presented in this volume. Recent research efforts should be summarized quantitatively to reflect present levels of benefits.

Fodder-fish integration practice in Malaysia

Raihan Sh. Hj. Ahmad

In Malaysia, integrated farming systems (Figure 1) have been practiced since the 1930s, with the production of fish in paddy fields and pig-fish in ponds. Although research shows that these systems are technically feasible and economically viable, socioeconomic factors, such as consumer preference, adoption by farmers, etc., need to be considered. Fodder-fish integration is one widely accepted system.

In the Third Malaysian Plan, fish culture is being promoted in a larger scope. Subsidies are given by the government for pond construction. Fish seed supply is provided as well as training and extension. This system benefits family consumption by providing enough supply of protein. Moreover, it can be a source of additional income.

The fodder-fish integration (Figure 2) utilizes the most commonly used fodder species as fish feeds (Figure 3). These are: napier grass *(Pennisetum purpureum)*, cassava *(Manihot esculenta)* and ipil-ipil *(Leucaena leucocephala)*.

Figure 1. Integrated farming system in Malaysia

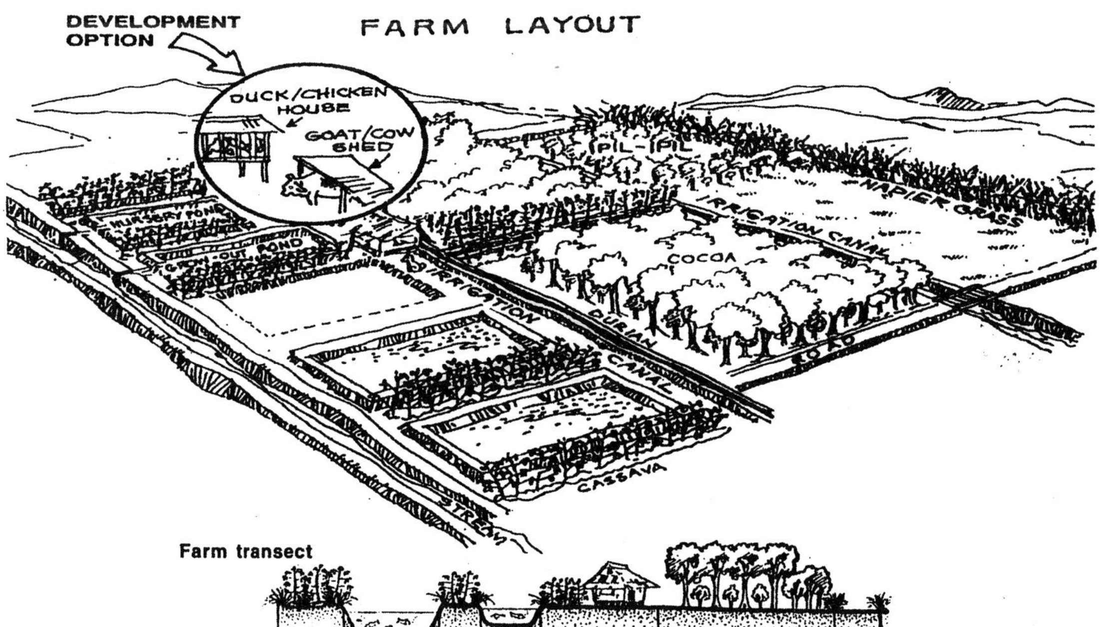

LAND TYPE	LOWLAND	LOWLAND	LOW-LAND	LOWLAND	LOW LAND	MEDIUM LAND	
SOIL	LOAM	CLAY LOAM	CLAY LOAM	CLAY LOAM	CLAY LOAM	LOAM	LOAM
WATER RESOURCES	RAIN	SMALL STREAM	RAIN	SMALL STREAM	RAIN	RAIN	RAIN
FODDER	NAPIER CASSAVA		NAPIER CASSAVA	DUCK WEEDS	NAPIER CASSAVA		NAPIER
TREES						IPIL-IPIL COCOA DURIAN	
FISH		• GRASS CARP • PUNTIUS • TILAPIA • BIGHEAD CARP • FRESHWATER PRAWN		GRASS CARP			

Figure 2. Calendar of activities for fodder-fish integration

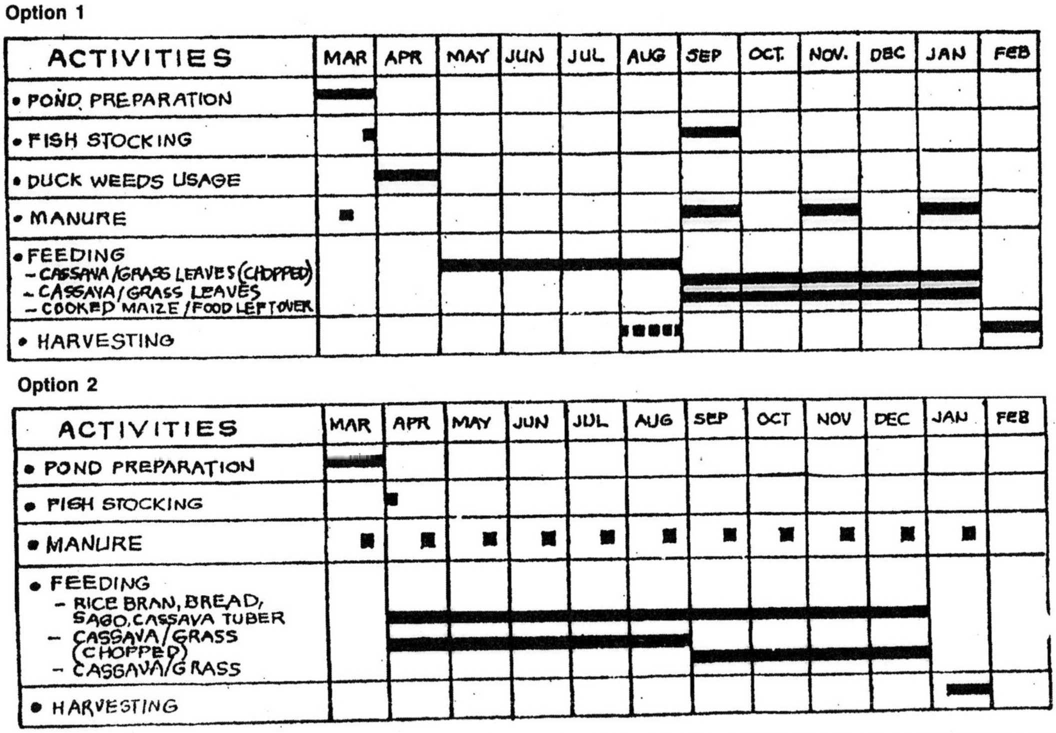

Figure 3. Fodder species used as fish feeds

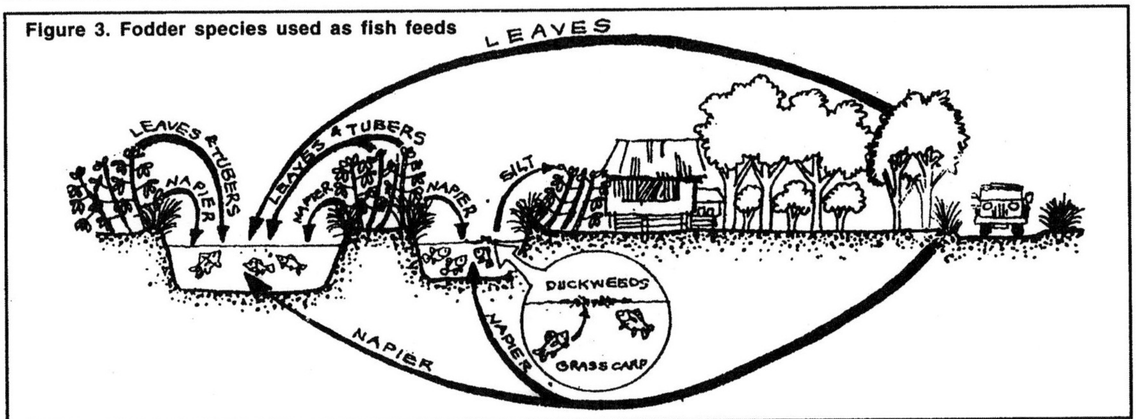

Land preparation and planting

1. Weed the land.
2. Plant fodder crops.
   - Napier grass and cassava are propagated by vegetative means using mature stems. Napier grass cuttings should have 3-5 nodes, three-fourth of which is buried (at about 45° angle). Cassava planting material is 25-30 cm long.
   - Ipil-ipil can be direct seeded or transplanted. Direct seeding is done when annual rainfall is 1 200 mm. Seedlings are best transplanted (at 2 cm depth) at the start of the rainy season.
3. Management care
   - If possible, put a fence around the area.
   - Do not allow grazing of animals.
   - Apply fertilizer/compost every month.
4. Harvest the fodder.
   - Napier: first cutting at 7 cm from the ground (to encourage vegetative growth) 6-8 weeks after planting. Then, cut regularly every 2-4 weeks, 10-15 cm from the ground.
   - Cassava: first cutting 0.5 m from the ground, 8 weeks after planting, then regularly after every 4 weeks.
   - Legumes: first cutting 8-12 months after planting, then regularly after every 8-12 weeks, 0.3 m from the ground.
5. Feed preparation
   - Leaves of these fodder crops are used as feeds. However, for cassava, the tuber can also be used. The leaves are chopped in small pieces before feeding to hatchlings or fry. For big fish, the leaves are simply placed in the pond.

Fish culture system

1. Pond design

The pond (0.1-0.5 ha in size) should be established near water sources and should be free from flood or drought.

Bunds are built to separate the ponds. Bunds should be between 2 and 3 m and capable of holding a water depth of 1 m. Water is supplied through gravity flow. Screened inlet and output pipes are installed.

A feeding area within the pond is constructed (located at the side). Bamboo poles or trunks of trees can be used.

There are two types of pond:

- Nursery pond - used for nursing 2.5-7.5 cm fry until the desired size is reached.
- Growout pond - bigger than the nursery pond, it is used to raise fish up to marketable size or to grow fish for breeding.

2. Pond preparation and system establishment

- Drain the pond (if the pond is an old one from which the fish have been harvested). Remove silt on the pond bottoms; this can be used as fertilizer.

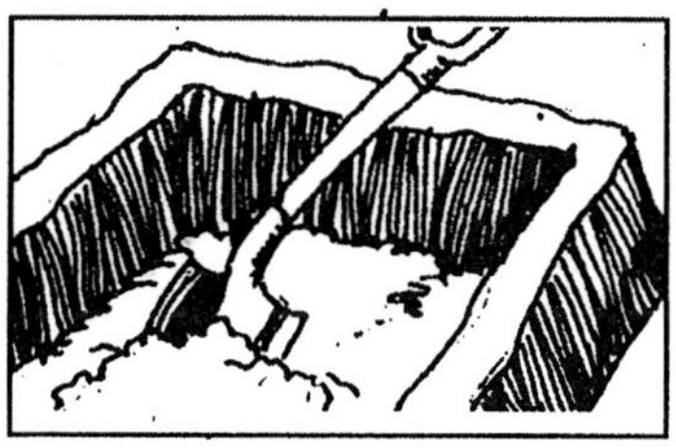

- Dry the pond bottom until the soil cracks. Plowing it first turns the soil over and facilitates drying.

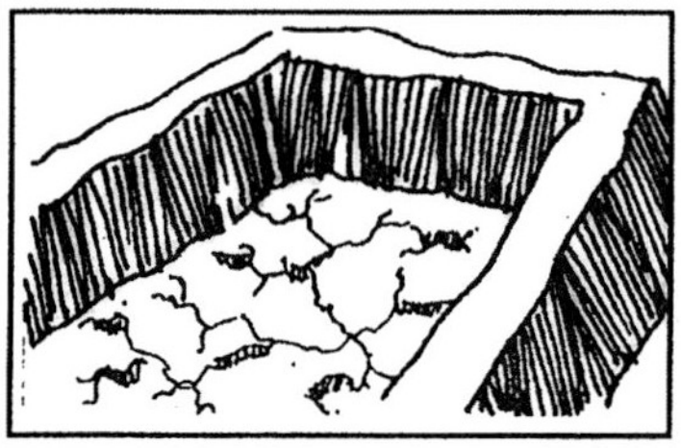

- Apply lime to condition the soil.

Liming activates fertilizers and controls acidic soils which may harm the fish. Quicklime is most commonly used at 200 kg/ha.

- Fill the pond with water 2 weeks after liming. Water should fall from the water inlet into the pond below, so that the water mixes with oxygen from the air. Also check water condition:
  - temperature = 22-32°C
  - early morning oxygen = 3 mg/litre
  - pH = 6.5-8.3

- Add fertilizer to the pond to provide nutrients for fish and plankton growth. Chicken manure can be applied at the following rates per hectare:

Option 1		
Nursery pond	:	200 kg (first month)
Growout pond	:	300 kg (first month)
		300 kg (third month)
		300 kg (fifth month)
Option 2		
Growout pond	:	100 kg (first month)
		20 kg (succeeding month)

- Stock the pond, preferably in the evening.

  Option 1: Grass carp is cultured in the nursery pond. After 4-6 months, the fish are transferred to the growout pond with the bighead carp and tilapia.
- Nursery pond (0.2 ha) - 500 pieces of grass carp (10 cm in size).

Growout pond (0.3 ha)	
Grass carp	500
Bighead carp	100 (1.5 cm)
Tilapia	1 500 (2.5 cm)

Option 2: Fish and prawn can be stocked directly to the growout pond.

Grass carp	100 (10-13 cm)
Javanese carp	300 (10-13 cm)
Giant freshwater prawn	3 000 (1 cm)

- Daily management of fishponds
  - Check the pond for leaks. Clean filters.

  If the fish are at the pond surface, feeds are needed. If they are gasping at the surface or the prawn are in the periphery of the pond, aeration is

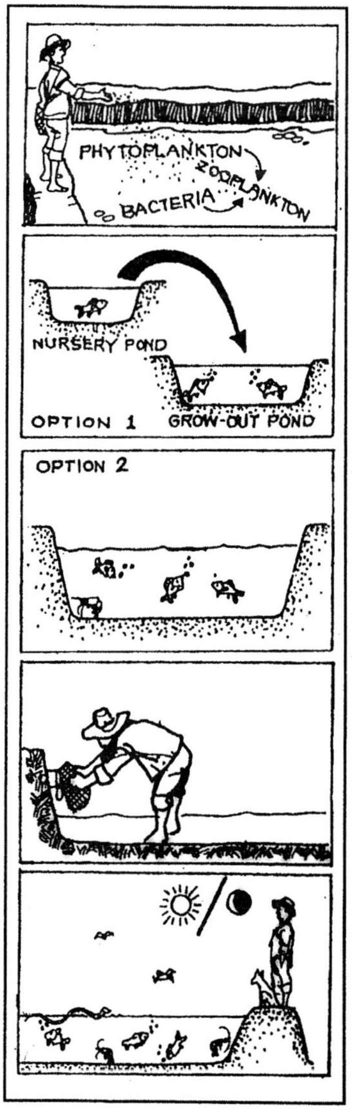

needed. Aerate the pond by stirring the water with a tree branch. Suspend fertilizer additions and later resume with reduced rates.

Also, watch for predators.

- Feed the fish/prawn.

Option 1: After the pond is fertilized, introduce duckweeds. Grass carp feed on duckweeds for the first month. Then, give chopped cassava leaves and napier grass. Feeding is twice a day (morning and afternoon). Upon transfer into the growout pond, feed the fish with grass and cassava leaves (200 kg/day). For tilapia, cooked maize, food leftovers and chopped cassava are given. The amount depends on the fish behaviour. If the fish are still in the feeding area, more feeds are needed.

Option 2: At the start, feed the

fish four times a day. Give rice bran, bread, chopped sago, cassava and napier grass.

For the fish, give feeds inside

the feeding area. For the prawn, broadcast the feeds all over the pond. If there are still feeds found in the water, stop feeding.

- Monthly management of fishpond
  - Check the pond walls and bottom. Remove any debris which might be a problem at harvest time, e.g. twigs,

leaves, etc.

  - Check the fertility and turbidity of the water by dipping your arm into the water. If the palm disappears before the water reaches the elbow, there is dense algal bloom.
  - Check the fish carefully for any sign of disease.
- About three to four partial harvests can be done using a sieve net before final harvest. For prawn, harvesting is after 6-7 months and for fish, after 10-12 months. Survival rate is about 70-90 percent for fish and about 30 percent for prawn.

3. Potentials
   - Environmentally sound
   - Prawn/fish is of high (economic) value
   - Seed is easily available.
   - With polyculture, different water columns are used, minimizing competition for food among different species.
   - Acceptable to consumers (as against fish grown in ponds loaded with manure or sewage)
   - The fodder crop can last for 5-7 years with minimum maintenance.
   - System is open to the introduction of additional components at a later stage.
   - Various combinations can be used to get highest yields and incomes.
4. Limitations
   - Cannot be applied on a large-scale basis
   - Requires high-labour inputs

Budget (in M$) for fodder-fish integration for 1 ha pond

Items	Option 1	Option 2
Costs		
Crops		
Land clearing, burning, soil preparation, fertilizing, seeding and maintenance of 1 ha of land	300	
Maintenance cost	150	150
Fish seeds		
Grass carp	225 (500 pcs)	40 (100 pcs)
Bighead carp	100 (100 pcs)	-
Tilapia	300 (1 500 pcs)	-
Javanese carp	-	15 (300 pcs)
Prawn	-	120 (3 000 pcs)
Feeds	300	30
Chicken manure	360	-
Lime	125	-
Miscellaneous	300	-
Total expenses	2 160	355
Income		
Grass carp	3 200 (800 kg)	180 (60 kg)
Bighead carp	510 (300 kg)	-
Tilapia	3 240 (720 kg)	-
Javanese carp	-	180 (60 kg)
Prawn	-	700 (70 kg)
Total income	6 950	1 060
Balance	4 790	705
Annual return (2 cycles/year)	9 580	-

1992: US$1 = M$2.70

Integrated fish-horticulture farming in India

S.D.Tripathi and B.K. Sharma

Integration of fruit and vegetable farming on the fishpond embankment has been tested in India, and has several advantages:

- The farmer gets additional income from growing fruits and vegetables on the pond embankment that normally lies fallow.
- The nutrient-rich pond mud is used as fertilizer for growing crops, eliminating the cost of organic manures.
- Manured pond water is used for irrigation of plants.
- Fruit and vegetable residues are used as feed for the fish.
- The plants on the embankment strengthen the dikes.

Fish-crop farming material flow

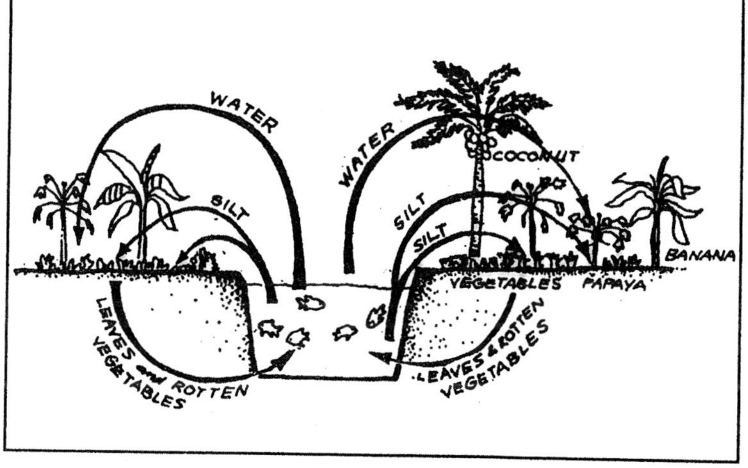

Fish culture

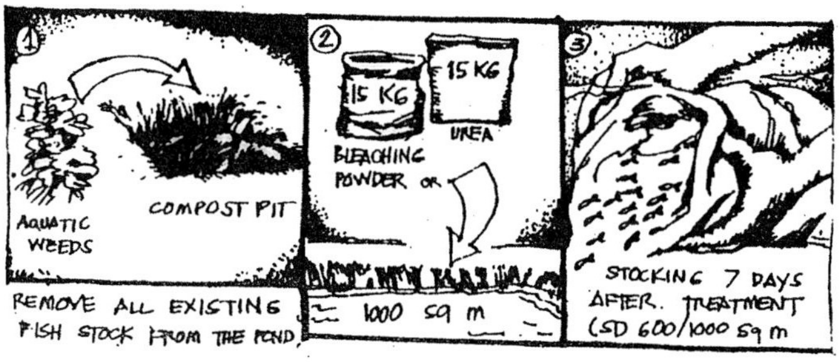

Establishment of the system

Select ponds near to your house. This helps in easy management of the pond and in discouraging poachers.

Check and repair the dikes and guard the inlets and outlets with meshed screens to avoid escape of stocked fish and entry of unwanted fish. The pond should be deep enough so that it retains more than 1 m water during the dry period.

Strengthen the dikes and terrace them for planting crops and fruit plants.

Pond preparation

Remove aquatic weeds. Compost and use them later as manure for the pond. Remove all existing fish stock from the pond by repeated netting and draining the pond water. If it is not possible to drain the pond, kill the fishes by adding to the water 15 kg bleaching powder and 15 kg urea (for 1 000 m^2 pond). Bleaching powder may be applied one day after urea application. Application of 250 kg Mahua oil cake *(Basia latifolia)* can also be done for the eradication of fish. Mix it thoroughly with the pond water and net all the fishes.

Manure the pond with the compost (made out of aquatic weeds). Apply 500 kg basally; the rest (500 kg) may be applied in two equal installments at 4 months interval but more frequent doses (e.g. fortnightly) are better.

Stock the pond with fingerlings 7 days after poisoning as the toxicity of bleaching powder lasts for about 1 week. The recommended rates (at stocking density of 600/1 000 m²) are:

Catla	240	Catla	180	Catla	90
Rohu	180	Rohu	180	Rohu	120
Mrigal	180	Mrigal	120	Mrigal	90
		Common carp	120	Silver carp	90
				Grass carp	90
				Common carp	120

Some alterations can be made on the stocking density and species ratio depending upon the pond conditions and availability of fish seed.

Harvesting

The fish that attain marketable size should be harvested and the rest allowed growing further. Final harvesting may be done 10-12 months after stocking.

Calendar of activities for fish-horticulture farming

August	Pond preparation
	Dike preparation and planting of fruits and vegetables
September	Stocking of fish
	Application of inorganic fertilizers to crops
October	Pest control if necessary
November	Harvesting of vegetables, inorganic fertilizer application
December	Harvesting of vegetables
January	Harvesting of vegetables
	Harvesting of papaya
February	Preparation of dike for second crop of vegetables
	Harvesting of papaya
	Plantation of second crop of vegetables
March	Partial harvesting of fish
	Harvesting of papaya and banana
April	Harvesting of papaya and banana
May	Harvesting of vegetables (P & B)
June	Harvesting of vegetables (P & B)
July	Final harvesting of fish
	Harvesting of vegetables, papaya and banana (P&B)

Horticulture

The dikes are strengthened, terraced, prepared and fertilized by application of pond silt.

Bananas, papayas, pumpkins, gourds, spinach, brinjals, tomatoes, cucumbers and leafy vegetables are grown on the dikes.

Inorganic fertilizer is also applied to the plants in addition to pond silt at 10 kg/year divided into installments.

Water the crops with manure pond water.

Planting of papaya is done in June/July and banana in October/November and harvesting starts after 6 and 8 months following planting, respectively. The farmer consumes a portion of the harvested fruits and the rest are sold in the market.

The vegetable crops are grown and harvested twice in a year—once during August/September and another in March/April. After meeting the requirements of the farm family, the vegetables are sold. Below is a list of some crops that can be grown on the pond embankment:

Fruit plants
- Papaya
- Banana
- Coconut

Vegetable plants
- Brinjal
- Cabbage
- Cauliflower
- Tomato
- Cucumber
- Pumpkin
- Ashgourd
- Bottle gourd
- Radish
- Beans
- Cowpea
- Okra
- *Colocasia*
- and other leafy vegetables

Budget (in rupee) for fish-horticulture in 0.1 ha pond

Cost	Rs
Pond preparation	
15 kg bleaching powder and 15 kg urea at Rs4.15/kg	125
Manuring with compost from aquatic weeds, 1 000 kg	100
600 fingerlings at Rs250/1 000	150
Labour and nets	200
Fish culture tools and equipment	25
Planting materials	
10 banana suckers at Rs2/pc	20
20 papaya seedlings at Rs1/pc	20
Vegetable seeds	6
10 kg inorganic fertilizers at Rs5/kg	50
Pesticides/horticulture equipment	25
Total cost	800
Pond rental (opportunity cost)	800
Interest on working capital at 5%	165
Total operational costs	1 265
Income	
Fish sale (200 kg at Rs20/kg)	4 000
Papaya (150 kg at Rs3/kg)	450
Banana (10 bunches at Rs20/bunch)	200
Green vegetables (159 kg at Rs3/kg)	450
Balance	3 835

Cash flow for integrated fish-horticulture farming for a 0.1 ha pond

	Aug	Sep	Oct	Nov	Dec	Jan	Feb	Mar	Apr	May	Jun	Jul
Inflow	0	0	0	+60	+90	+75	+75	+1 075	+115	+160	+205	+3 245
Outflow	-314	-199	-64	-49	-44	-49	-94	-179	-44	-49	-44	136

Notes:
1. Cash inflow from November when harvesting of vegetable is initiated
2. Cash flow goes up to Rs1 075 in March, when partial harvesting of fish is also done
3. Cash inflow includes harvesting of second crop of vegetables from April to July

Issues for further consideration

In India, semi-intensive commercial systems are well established in certain areas. However, the adoption of similar technologies among resource-poor farmers is not well documented, and the system described here is a downscaled version of the semi-intensive commercial system. In practice, downscaling operational procedures and rates are difficult to implement as resource-poor farmers cannot access the materials in the required amounts, frequency and quality (e.g. lime, manure, weeds) or in the exact stocking ratios and sizes for polycultures, etc. In many situations, smallholder ponds are undrainable and cannot be dried, so pond mud removal is difficult and laborious, and pumping costs can be prohibitive.

Numerous other examples of vegetable-fish and fruit-fish integration from other countries, farming systems and agroecological zones exist in the scientific literature. The example presented here should encourage prospective users to consider, given their own resources and agroecological conditions, which vegetable crops are suitable and marketable. Further, they need to decide at which location on the pond dikes the crops should be planted, based on seasonality requirements for shade and water, etc. Usually the availability of land and labour on the farm, together with household and market needs, will determine if farmers could embark on these activities. Supporting institutions would have to assess the economic viability of the new aquaculture component and its synergistic benefits in comparison to other opportunities.

Culture of short-cycle species in seasonal ponds and ditches in Bangladesh

Modadugu V. Gupta

Homestead seasonal ponds, ditches and road-side canals, which are formed either due to burrowing of soil for house or road construction or ponds dug for household uses (bathing, washing) or irrigation, can be used for aquaculture of short-cycle species, such as silver barb (*Puntius gonionotus)* or Nile tilapia *(Oreochromis niloticus).* Even 80-100 m² ditches as shallow as 70-80 cm can be used for culture of these species, using on-farm agricultural wastes and by-products, as inputs. Even ponds, which retain water for only 3-4 months, can be used for culture of these species. The culture practice is simple, requiring very low labour input and, hence, can be undertaken by women and children, producing fish for household consumption and for market. Landless farmers can also benefit from this technology by

Agroecosystem transect of Mymensingh-Sylhet floodplain area in Bangladesh

ROAD LAND TYPE	ROAD SIDE POND	HOMESTEAD	HOMESTEAD UPLAND	MEDIUM LAND	LOWLAND	FLOOD PLAIN	RIVER
SOIL	SANDY LOAM			LOAM	SILTY CLAY LOAM	CLAY	LOAM
CROPS	JUTE RET-TING		RICE, PULSE, PEAGRAM, BEANS, WHEAT, TOMATO, POTATO, GARLIC, MUSTARD, ONION, CABBAGE, LADIES FINGER	RICE WHEAT GRAM SUGARCANE	RICE WHEAT GRAM	RICE	
TREES		MANGO, LITCHI, GUAVA, PAPAYA, DATE PALM, COCONUT					
LIVE-STOCK		GOAT, CATTLE, BUFFALO, POULTRY					
FISH	WILD FISH, SILVER BARB, TILAPIA			CARP, TILAPIA, SILVER BARB, WILD FISH		FISH (SEASON-AL)	WILD FISH

MATERIAL FLOW BETWEEN HOMESTEAD ENTERPRISES
(landless farmer only - left portion; crop farmer - middle and right portion) **IN BANGLADESH**

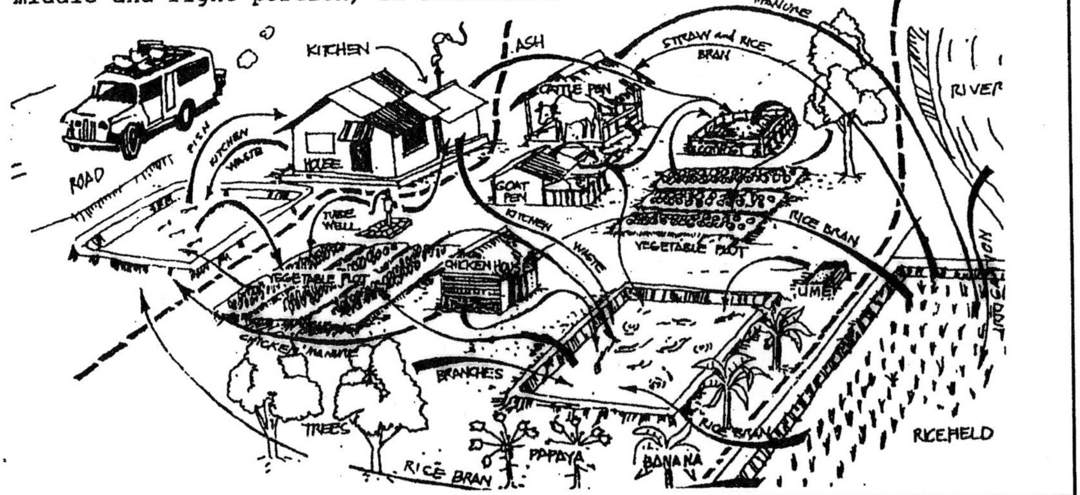

culturing fish in common property roadside ditches

Technology for culture of short-cycle species

1. Pond preparation
   - Branches of trees on pond embankment should be cut or trimmed. Pond should be cleared of submerged and floating weeds as they utilize pond nutrients and obstruct penetration of sunlight into water, resulting in low production of fish food organisms.
     - Clear away weeds.
     - Trim branches to allow sunlight.
   - For lowering of acidity, better utilization of fertilizer and disinfection, apply lime to the pond at the rate of 25 g/m^2. Spread lime on the bottom if pond is dry, or dissolve in water and spray if pond is with water.

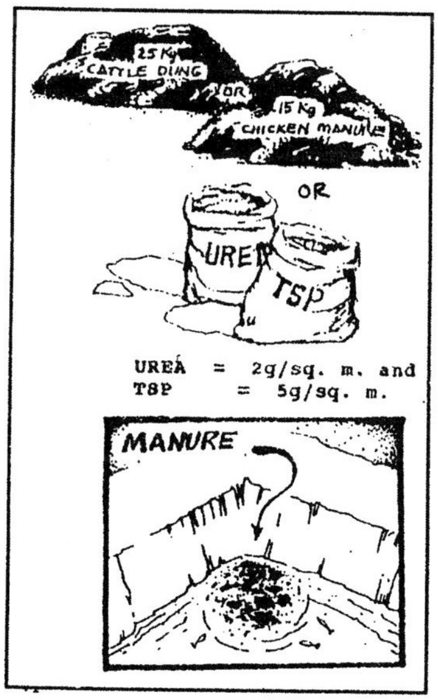

2. Fertilization
   - For good production of fish food organisms (plankton) on which fish growth depends, the pond needs to be fertilized. Organic manures and/or chemical fertilizers can be used for the purpose. Cattle dung (100 g/m^2) or chicken manure (50 g/m^2) or urea (2 g/m^2) and triplesuperphosphate (5 g/m^2) need to be applied once every 2 weeks.
   - Organic manure can be heaped in the corners of the pond while chemical fertilizers need to be dissolved in water and spread in the pond.
   - Every 2 weeks, a pond of 500 m^2 needs 25 kg of cattle dung or 15 kg of chicken manure or 1 kg of urea and 2.5 kg triplesuperphosphate. Schedules can be more frequent, e.g. weekly or even daily.

3. Stocking
   - Depending on farmer's choice, Nile tilapia or silver barb (*P. gonionotus*) can be cultured in the pond.
   - In case of Nile tilapia, 2 fingerlings/m^2, while in case of silver barb, 3 fingerlings/ m^2 need to be stocked.
   - If the pond retains water for more than 6 months, in addition to silver barb, 3 fingerlings/40 m^2 of catla *(Catla catla)* or silver carp *(Hypophthalmichthys molitrix)* and 2 fingerlings/40 m^2 of common carp *(Cyprinus carpio)* need to be stocked. This will increase total fish production.
   - Healthy fingerlings should be procured from a reliable hatchery or supplier. It is better to stock 3-5 g size, as they would reach table size early, especially in case of ponds which retain water for only 3-4 months.

4. Feeding
   - For good production, supplementary feeds should be given in the pond. Kitchen waste, duckweeds, azolla, kangkong leaves, sweet potato and tender terrestrial grasses can be given. Rice bran or wheat bran will increase growth and production of fish.
   - Feeding should be done once or twice a day. Quantity to be given increases with size of fish. A tentative schedule of feeding rice bran in a 500 m^2 pond is shown in chart. If kitchen wastes or weeds are given, quantity of rice bran shown in chart can be reduced.

5. Pond management
   - Green color of water indicates good production of fish food organisms (plankton). Clear water indicates lack of enough fish food. By dipping your hand in

the water, seeing it halfway to the elbow indicates lack of enough fish food. In such case, increase fertilization. If the hand disappears halfway to elbow, this indicates sufficient plankton in the pond. If the palm disappears after dipping, this indicates plankton bloom, with deep green color, and this can deplete oxygen in pond water, especially during night and cloudy days and can result in mortality of stocked fish. Feeding and fertilization should be stopped until the water color becomes lighter.

- Tilapia breeds in pond, leading to overpopulation. This results in poor growth of fish due to competition for food. Hence, tilapia fry that move in schools along the banks of pond can be removed, using a scoop net. They can either be sold or crushed and given as feed in the pond.

6. Harvesting
   - Harvesting of fish can be started as soon as they reach table size or when the water level in the pond goes below 40-50 cm. You may harvest for family consumption or at one time for marketing. About 75-100 kg of fish could be harvested from a 500 m² pond in 5-6 months.

7. Disease
   - When temperature goes down to about 20°C and below during November-January, silver barb is susceptible to ulcerative syndrome disease, which starts as red spots on fish and later becomes a wound.
   - When infection is seen, apply lime in pond at the rate of 25 g/m² pond areas.

Budget (in taka) for *Puntius gonionotus* and Nile tilapia culture in a seasonal pond of 500 m² for 6 months

Costs	*Puntius*	*Tilapia*
300 kg cattle dung at TK.35/kg	105	105
300 kg rice bran at TK1.50/kg	450	450
Labour for pond cleaning and harvesting	135	135
12.5 kg lime at TK3/kg	38	38
750 fingerlings at TK.30 each	225	250
Transportation cost of fingerlings	30	30
Total costs	983	1 008
Income		
75 kg fish at TK40/kg	3 000	2 250
Balance	2 017	1 242

1992: US$1 = TK38
Note: If on-farm sources of cattle dung and rice bran are used along with family labour, then TK690 can be saved which will raise the balance to TK2 707 for *Puntius* and TK1 932 for *Tilapia*.

Issues for further consideration

This culture is widely practiced in the area described and is an important system for poorer people. However, it remains to be clarified if landless people take control of their local common property and intensify it in the way described, using disinfection and fertilization. Social aspects are a critical issue. Multiple ownership of ponds is a common constraint to increased levels of management, and often leads to erratic or infrequent management.

This practice is perceived to have high potential, but in order to formulate recommendations and estimate potential areas and characteristics and numbers of likely adopters, critical information needs to be documented: e.g. the mode in which small communities or extended family groups have organized themselves; the extent to which the procedure has already been adopted; the range of benefits it has actually brought to households, etc.

In respect to "table-size" fish, recent research in northwest Bangladesh indicates that small tilapia are acceptable to resource-poor households and are even desired by women as they have little cash value and are therefore more likely to be consumed by all family members.

The importance of these systems to marginal people in the community (e.g. the landless) and in the household (e.g. women) is noteworthy. Recent results show that access by women and the system's productivity can be enhanced with the right extension approach.

ANIMAL-FISH SYSTEMS

Integrated fish-duck farming

S.D. Tripathi and B.K. Sharma

Raising ducks over fishponds fits very well with the fish polyculture system, as the ducks are highly compatible with cultivated fishes. The system is advantageous to farmers in many ways:

1. Ducks fertilize the pond by their droppings when given free range over the pond surface. Ducks have been termed as manuring machines for their efficient and labour-saving method of pond manuring, resulting in complete savings on pond fertilizer and supplementary fish feed which accounts for 60 percent of the total cost in conventional fish culture.
2. Ducks keep water plants in check.
3. Ducks loosen the pond bottom with their dabbling and help in release of nutrients from the soil, which increase pond productivity.
4. Ducks aerate the water while swimming; thus, they have been called "biological aerators."
5. Duck houses are constructed on pond dikes; hence, no additional land is required for duckery activities.
6. Ducks get most of their total feed requirements from the pond in the form of aquatic weeds, insects, larvae, earthworms, etc. They need very little feed, and farmers normally give kitchen wastes, molasses and rice bran, for the purpose.

Cultural practices

Successful pond management is the basis of profitable fish culture. Build the pond (about 1 000 m²) near your house to enable you to take proper care of your ducks and fish and to discourage poaching.

Check the pond dikes and repair the damages, if any. Deepen the pond so that it retains more than 1 m depth during the dry season.

Drain or dry the pond and remove or kill all the remaining fish stock from the pond by applying 15 kg bleaching powder and 15 kg urea/1 000 m².

Urea and bleaching powder may be applied one after the other and the dead fish netted out. Alternatively, 250 kg of Mahua oil cake *(Basia latifolia)* may be applied which not only kills fish but also acts as pond fertilizer.

Apply 20-25 kg of lime about a week before manuring the pond. In case a mixture of bleaching powder and urea is applied to eradicate the predatory and weed fishes, apply only 5-10 kg of lime (reducing the amount of bleaching powder applied).

Manure the pond with a basal dose of cattle dung at 500 kg/1 000 m². Stock the pond with fingerlings 7 days after poisoning as the toxicity of bleaching powder lasts for about 1 week. The fingerlings of over 10 cm size should be stocked, as the smaller ones are likely to be preyed upon by the ducks. The recommended rates of stocking (per 1 000 m² at a stocking density of 600 pieces) are shown on the following table.

Some alterations can be made on the stocking density and species ratio depending upon the pond conditions and availability of fish seed.

Grass carp should be fed regularly with aquatic or terrestrial vegetations. It should be fed before the ducks are allowed to come out of the duck house; otherwise, they will spread the weeds over the entire pond surface.

Culture	Species		Number
3 species culture			
	Catla	-	240
	Rohu	-	180
	Mrigal	-	180
4 species culture			
	Catla	-	180
	Rohu	-	180
	Mrigal	-	120
	Common carp	-	120
6 species culture			
	Catla	-	90
	Rohu	-	120
	Mrigal	-	90
	Silver carp	-	90
	Grass carp	-	90
	Common carp	-	120

6 carp species

Harvesting

Fish which attain marketable size should be harvested and the rest allowed to grow further.

Final harvesting may be done 10-12 months after stocking.

CALENDAR FOR DUCK-FISH-DUCK FARMING

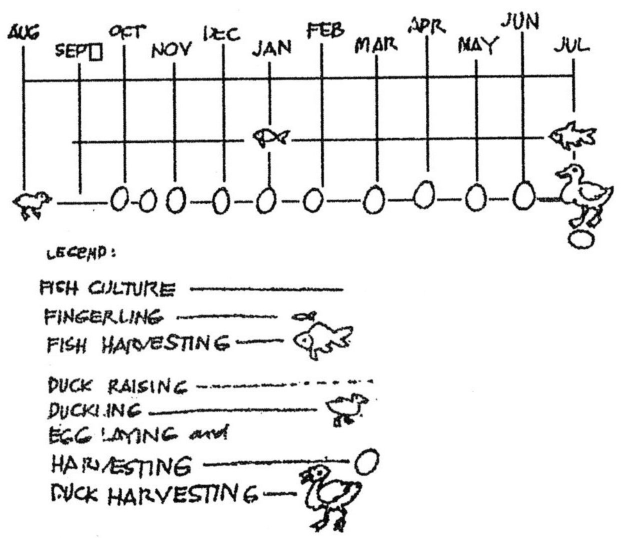

Duck farming

Egg laying by ducks depends upon many factors, including breed and strain, but good management contributes considerably towards the achievement of optimum egg-flesh production.

The ducks do not need elaborate housing since they remain in the pond most of the day. A low-cost night shaker made of

bamboo or any other cheap material should be available in the area either on the pond embankment or on the water surface. The house should be well-ventilated and so designed that the washings are drained into the pond.

About 30 ducks are sufficient to fertilize a pond of 1 000 m²; this number only needs a (house) floor area of 13 to 14 m². About 3-4-month old ducklings are kept on the pond after giving them necessary prophylate treatment and safeguarding measures against epidemics.

The ducks can find natural food from the pond. They will need very little supplementary feed which can come from household wastes, such as kitchen leftovers, rice bran, broken rice and spoiled cereals, if any. Alternatively, a balanced feed may be purchased and given at 50 g/bird/day.

Budget (in rupee) for fish-duck farming in 0.1 ha pond

	Rs
Costs	
Pond preparation with 15 kg bleaching powder (or 5-10 kg lime) and 5 kg urea at Rs4.15/kg	125
Manuring with basal dose of 500 kg cattle dung at Rs10/100 kg	50
600 fingerlings at Rs250/1 000	150
Net and labour for harvesting	300
Fish culture tools	25
	650
Bamboo duck house (minimal cost)	300
30 ducklings (4 months old at Rs20/bird)	600
810 kg feed at Rs3/kg	2 430
Medicines	20
	3 350
Total costs	4 000
Income	
Fish sale (400 kg at Rs20/kg)	8 000
Egg sale (3 100 at Rs100/100)	3 100
Duck sale (60 kg at Rs20/kg)	1 200
Total income	12 300
Balance	8 300

Fish-duck farming material flow

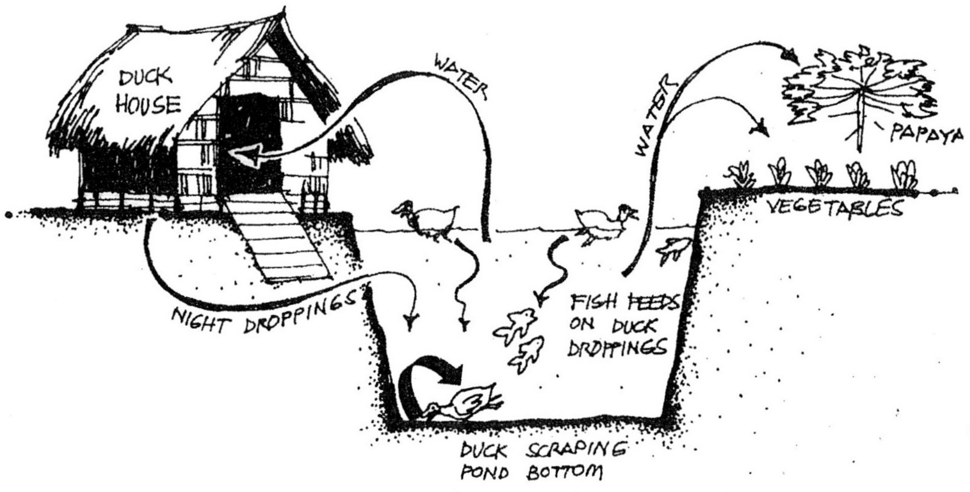

Cash flow for integrated fish-duck farming for 0.1 ha pond

	Aug	Sep	Oct	Nov	Dec	Jan	Feb	Mar	Apr	May	Jun	Jul
Inflow	0	0	+225	+225	+300	+300	+2 300	+300	+2 350	+350	+350	+5 550
Outflow	-1 650	-406	-256	-256	-256	-256	-356	-256	-356	-256	-256	-377
Netflow	-1 658	-2 064	-2 095	-2 076	-2 032	-1 988	-44	0	+1 994	+2 082	+2 182	7 355

Notes:
1. Cash inflow starts in October when the ducks begin laying.
2. Cash inflow increases in February and April when partial harvesting of fish is done.

Moldy feed, or feed kept for long time, should be avoided as molds contain toxins which may cause poisoning.

The ducks start laying at the age of 24 weeks. Laying boxes with straw may be kept in the duck house.

Proper sanitation and health care are very important to maintain a healthy stock. A sick bird is easy to detect: it becomes restless, its eyes lack brightness, and watery discharge comes out of the eyes and nostrils. The sick bird should immediately be isolated and treated.

The eggs are collected every morning as the ducks lay eggs only at night. The ducks lay eggs for two years, after which they should be culled.

Issues for further consideration

General: This case study is from an experimental system characterised by (a) complex polycultures; (b) relatively long fish culture periods of 10 to 12 months; (c) conventional, textbook pond preparation and pond management and the necessary inputs. This system has not been widely adopted by resource-poor farmers. The technology may be suitable only in areas where there is strong demand for duck eggs and duck meat.

Fish component: To produce fish yields of 4 t/ha/12 months of carp polyculture, a higher level of inputs will probably be necessary than just the droppings of 30 ducks.

Duck component: The user will be required to obtain further details of key features of managing ducks on fishponds which will be necessary to run a successful duck-fish growing activity. Among these, in an enclosed system, is the importance of ensuring that all duck waste enters the pond. This is expensive, because of the necessity to give complete feed to the ducks, in comparison to the less costly extensive system, in which ducks must be allowed to scavenge around and beyond the pond for natural food that is supplemented. Inevitably, a proportion of the duck wastes are lost as they roam beyond the pond. Thirty ducks may not be able to find most of their food from a pond of 1 000 m^2. Most strains of ducks should not be raised for more than 12 months as their productivity declines. Other experiences have shown that chopped water hyacinth is a poor feed for ducks.

Economics: The cash flow analysis will be different for local applications and conditions, and will require adaptive trials. Overall, the technology may be viable only under certain operating conditions. The duck rearing component may be economically viable only as long as the ducks are fed exclusively with on-farm wastes and kitchen refuse. This restricts the number of ducks to be kept in the system, usually five to ten sustained by wastes of the single household. Once the farmer has to depend on purchased feed (as far as this is even available in rural areas) an economically viable size of the duck farm has to be maintained, which will depend on local preferences, market conditions, egg and meat prices, and cost of duck feed. From past experience, rearing units with even 100 to 200 ducks could not be sustained for long.

In the described case and from the farmer's point-of-view, it will be necessary to spend an amount of Indian Rs 3 350 to obtain a marginal profit of about Rs 995 from duck farming. On the other hand, it is possible to get Rs 7 000 by spending about Rs 700 (including the cost of lime) from aquaculture. The farmer in such case would prefer to secure and apply organic manure from other sources than to take up duck farming where such a high investment and risk are involved. It is because of these factors that this integrated fish farming technology package in its present form has not become popular among resource-poor farmers in India, Bangladesh, Nepal and other neighbouring countries in spite of repeated efforts to popularize it over the last 10 to 15 years.

Integrated chicken-fish farming

Modadugu V. Gupta and Francisco Noble

Chicken raising for meat (broilers) or eggs (layers) can be integrated with fish culture to reduce costs on fertilizers and feeds in fish culture and maximize benefits. Chicken can be raised over or adjacent to the ponds and the poultry excreta recycled to fertilize the fishponds. Raising chickens over the pond has certain advantages: it maximizes the use of space; saves labour in transporting manure to the ponds and the poultry house is more hygienic. No significant differences have been observed on the chickens' growth or egg laying when they are raised over the ponds or on land. In case of the former, the pond embankment could still be utilized for raising vegetables.

Fish culture

1. Pond preparation
   - For an example pond of 1 000 m^2, remove predatory and weed fish either by draining the pond or by applying an approved piscicide.
   - Apply 25 kg lime to pond bottom if the pond is dry, or dissolve lime in water and spray solution if pond has water.
2. Stocking
   - Stock 600-1 000 fingerlings of Indian carps, catla *(Catla catla)*, rohu *(Labeo rohita)*, mrigal *(Cirrhinus mrigala)* and Chinese carps, silver carp *(Hypophthalmichthys molitrix)*, grass carp *(Ctenopharyngodon idella)* and common carp *(Cyprinus carpio)*. Species stocking rate could be 40 percent surface feeders (catla and silver carp), 20 percent rohu, 30 percent bottom feeders (mrigal and common carp) and 10 percent grass carp.
3. Feeding
   - No feeds need to be given, as the feed spilled by chicken (which could be as much as 10 percent) fall into ponds.
4. Fertilization
   - No fertilizer is needed, except for excreta of chicken falling into ponds.
5. Harvesting
   - Harvesting of fish could start 6-7 months after stocking when some fish reach table size.
6. Oxygen depletion
   - When water becomes deep green due to plankton blooms, oxygen in the water may get depleted and fish may die. In such cases, put mats or plastic sheets below the poultry house to catch the chicken excreta and suspend nutrient inputs for 1 to

3 weeks. If possible, immediately irrigate the pond with freshwater.

Chicken raising

For the first 14 days, chicks need to be raised separately in a brooder (not on pond), as they need higher temperature of 28-33 °C (85-95 °F). Each chick during this period needs a space of 7.5 x 7.5 cm (9 in²). To maintain the required temperature range, surround the chicks in a bamboo fence and hang an electric or kerosene lamp above them. A rice husk heater can also be used.

1. Broilers: 30-50 broilers could be raised on a 1 000 m² pond.
   - The chicken house can be constructed over the pond at least 0.5 m above maximum pond water level, or on the embankment. Each bird requires 1.5 ft² space (50 birds require 75 ft² space). The house can be made of bamboo or any other locally available cheap materials. Roof can be covered with hay or similar material. Enough cross ventilation should be maintained to keep cool during hot days. Floors are to be constructed with bamboo slats, with 1 cm gap, to allow excreta to fall into pond, but not wide enough for the chicken's feet to get caught in between and injured.
   - Any fast-growing chicken, like Shavar Starbro broilers, can be raised.
   - Feed with starter mash for 1-4 weeks and with finisher mash for 5-8 weeks, given as much as they can consume. A 100-kg starter mash requires 50 kg crushed wheat, 14.5 kg rice bran, 16 kg sesame oil cake, 19 kg fishmeal and 0.5 kg salt. A 100 kg finisher mash requires 50 kg crushed wheat, 17 kg rice bran, 15 kg sesame oil cake, 16 kg fish meal, 1.5 kg bone meal and 0.5 kg salt. In both cases, vitamin premix is added at the rate of 250 g/100 kg of feed.

CHICKS ARE ASSURED OF WARMTH AS THEY ARE FENCED BY A BAMBOO ENCLOSURE.

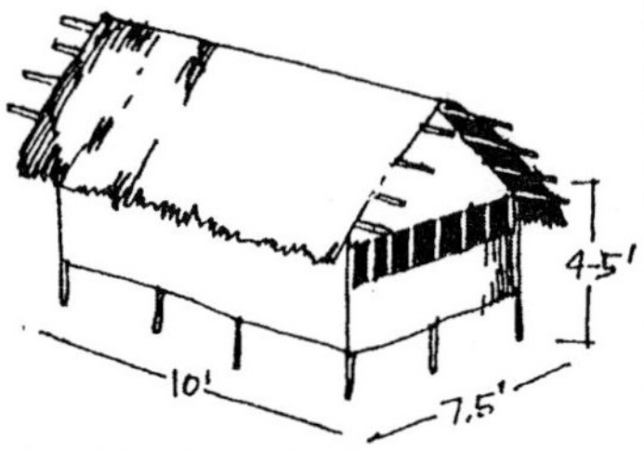

HOUSE FOR 50 BROILER CHICKENS

   - Water should be provided at all times.
2. Chicken layers: 30-50 layers can be raised over a pcnd of 1 000 m².
   - Housing can be constructed on pond or on embankment. Each bird requires 3 ft² floor area.
   - Any good strain of chicken, like Star Cross Shavar, could be raised.
   - For the first 16 weeks, feed is given at the rate of 80-110 g/bird/day and from 17th week onwards, 110-120 g/bird/day. Feed composition is given in the chart below.
   - Temperature in the poultry house should always be above 20-22°C. When the temperature goes below this level, hang two 200-watt bulbs or two kerosene lamps for every 50 chickens. A rice husk heater can also be used.
   - Broilers reach market size of 1.5-1.8 kg in 7-8 weeks and it is possible to raise six batches in a year. Layers start laying after 22 weeks and 250-280 eggs/bird/year could be obtained. Egg laying becomes uneconomical after chickens reach the age of 18 months, when they need to be replaced.
   - Because chickens are usually kept in confinement, they are susceptible to disease. When disease strikes, the whole flock may be affected: growth will be retarded, egg production will go down or the chickens may die. Thus, broilers will not reach market weight in time. For layers, sexual maturity is delayed. Protective measures are needed.

Vaccinate your chickens

In some countries, vaccines can be obtained from the nearest livestock office, free of cost. The following are some reminders when collecting vaccines:

Feed composition for layers

Ingredients	0-4 weeks	4-16 weeks	16 weeks above
Crushed wheat	46.0	46.0	44.0
Rice polish	20.5	24.5	29.0
Sesame oil cake	16.0	14.0	12.0
Fish meal	17.0	15.0	10.0
Bone meal	-	-	4.5
Salt	0.5	0.5	0.5

Vitamins premix at 250 g/100 kg feed.

- Bring a good thermoflask and a little cotton wool.
- Do not waste vaccine. Obtain only the exact amount needed. Vaccine production costs a lot to the government.
- Store vaccines at low temperature, preferably in a refrigerator, to maintain their effectiveness.

Equipment necessary in vaccination

- Thermoflask of sufficient capacity to carry the vaccines.
- Nylon syringe - one or two, graduated at 1 ml intervals. Smaller-capacity syringe is preferable.
- Needles of gauge 20 or 21 and 14 or 15. Shorter needles of 1-2 cm length are preferable for poultry vaccination. A few large sewing needles are suitably modified for fowl pox vaccination.
- Measuring cylinder
- Two wide-mouth bottles: one to carry distilled water and another to dilute vaccines, when necessary. These items preferably should be of nylon or polypropylene which could be sterilized by boiling when necessary.

Reminders when vaccinating chickens

- Sterilize syringes, needles and all other equipment before using.
- Put ice cubes at the bottom of the thermoflask and a layer of cotton wool before placing the vaccine vial. Close the flask.
- Check vaccine if it looks all right. Do not use discolored or unusual-looking vaccines.
- Use distilled water purchased from a pharmacy when diluting vaccines, or boil clean water for 10-15 minutes. Cool down, then strain into a clean bottle.
- When only a small amount of distilled water has to be added, draw the required amount into the sterile syringe and inject into the vial. Dissolve by vigorously shaking the vial.
- Pour the balanced amount of distilled water into the mixing bottle. Draw the dissolved vaccine into the syringe. Pour into the mixing bottle containing the balanced quantity of distilled water. Thoroughly mix with a sterilized rod.
- In case of fowl pox, remove the required amount into a sterilized empty vial and use for vaccination. This prevents contamination and subsequent waste of surplus vaccine.
- Do not spill vaccines. This could be fatal to chickens.
- Hold the needle with the knob. Do not touch the tip when assembling the syringe for vaccination. Contaminated needles should not be used until sterilized.
- Before vaccination, confine the birds, picking up one by one and releasing after vaccination. This makes vaccination easy and no bird is missed.
- Do not vaccinate birds suffering from disease or in a state of stress. Vaccinate them

Chicken disease and preventive treatment

Vaccination technique	Medical dosage	Consequences if you do not vaccinate
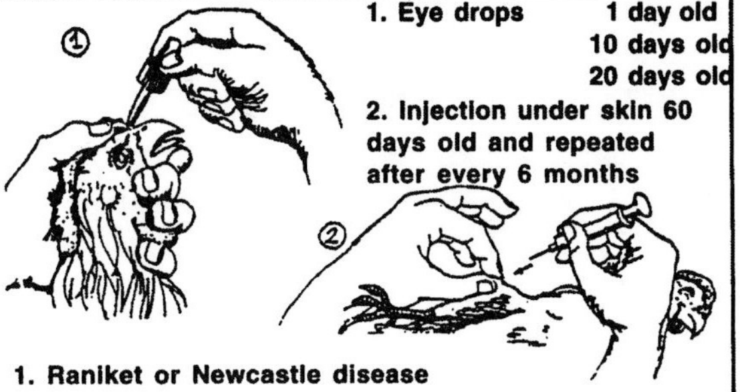 1. Eye drops 1 day old 10 days old 20 days old 2. Injection under skin 60 days old and repeated after every 6 months 1. Raniket or Newcastle disease	BCRDV 1 drop into the eye RDV 1 ml into the thigh	Show difficulty in breathing, unusual gait, moving in circles, walking in backward direction, head hidden between legs. 
Pricking under wing bed 30 days old and repeated after every 6 months.  2. Fowl pox	FP Vaccine 1 ml	 Pox lesions - small vesicles in the face comb and wattles
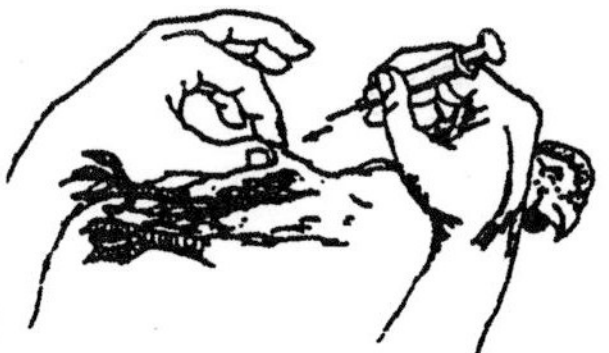 3. Fowl cholera Injection under skin 90 days old and repeated after every six months	1 ml	 Birds suffer from diarrhea and faeces is yellowish or greenish. The birds become droopy and sleepy; head will be drawn down or turned backward resting on the wing. Breathing is difficult.

only when they are back to normal.

- Two vaccines should not be given the same day. A 10-day gap is needed between two successive vaccinations.
- Record data so that the next vaccination will be known.
- If leftover diluted vaccine can be used within a short period, it should be put in a clean polythene bag and placed in the flask containing ice.
- Wash all equipment used with soap and clean water, then sterilize in boiling water.
- Thoroughly clean empty vaccine vials. Return them to the Livestock Officer when collecting the next requirement of vaccines.
- Vaccinate birds on time.

Budget (in taka) for a 1 000 m² pond in integrated chicken-fish farming with 50 broilers

Costs	
6 batches of broilers	
(1 batch is 53 days TK15.00/chick, 6% mortality)	4 770
Feed (202.5 kg at TK9.75/kg)	11 844
Vaccines	90
Fuel	240
Transportation	180
Lime (25 kg at TK3/kg)	75
Fish fingerlings (600 at TK400/1 000)	240
Labour for netting	200
Total costs	17 639
Income	
Broiler meat (6 batches x 50 chicks x 1.5 kg x TK52)	23 400
Fish (500 kg at TK35/kg)	17 500
Total income	40 900
Balance	23 261

Costs

- A broiler farmer needs capital for investment for raising one cycle of broilers, which he can sell after 7-8 weeks. The sale proceeds can be used for the next batch.
- A layer farmer needs capital investment for chicken feed for the first 22 weeks. When the hens start laying, the farmer can use the sale proceeds from eggs for chicken feed.
- A broiler farmer needs capital for:
  - Chicken shed: TK5 000; life expectancy, 3 years
  - Feeder and waterer: TK200; life expectancy, 4 years

Budget (In taka) for a 1 000 m² pond in integrated chicken-fish farming with 50 layers over an 18-month period

Costs	
52 day old chicks (at TK22 each, 4% mortality)	1 144
Feed	
8.5 kg/bird at TK9 up to 20 weeks	3 825
120 bird/day at TK7.75, 20-72 weeks	16 926
Vaccine	200
Fuel	500
Lime (25 kg at TK3/kg)	75
600 fish fingerlings at TK400/1 000	240
Labour for netting	200
Total costs	23 110
Income	
Chicken eggs (65% average production, 52 weeks x 7 days x 32.5 eggs x TK2.4)	28 392
Chicken (50 birds x 2.2 kg x TK45)	4 950
Fish (600 kg at TK35/kg)	21 000
Total income	54 342
Balance	31 232

Issues for further consideration

The case presented is a resource-intensive experimental system. Most farmers may find it difficult to sustain the 30-50 birds that are recommended. This activity may be adopted by somewhat "better off" farmers with an entrepreneurial focus. Identifying who adopted this in existing communities and who could be potential adopters in other communities could be done through ranking exercises.

Although the importance of chicken diseases is emphasized in the example, most resource-poor farmers are at least equally constrained by the lack of, or poor quality of existing, chicken feed. In implementing the technology, more information will be required on what proportion of the feed can be produced by the farm household, or is produced and purchased locally, or needs to be supplied from outside the village (e.g. fishmeal, bonemeal, salt, vitamins). Also, information on who carries out the activities and who benefits within the household will be useful for planning.

From experience with this technology in field trials, the poultry component was the problematic part. It required a higher level of investment and managerial skill. Uneconomical scales of chicken farming operations limited acceptance by farmers and caused the few adopters to terminate the activities.

Limited levels of integration can be achieved with backyard chicken rearing. Farmers usually rear few birds (usually only 5-10) and let them roam freely, avoiding the need for purchasing feed.

Any development of a new vaccination program in a particular area would need to take the experience of farmers and the local livestock office on the most important diseases in the region into account. In addition to the preventive vaccination the need for curative treatment, e.g. of parasitic diseases may arise.

The procedure for vaccination (amount, mode of application etc.) may vary between different brands, and in this respect the guidelines given by the vaccine producer should allways be followed. Tap water containing chlorine can inactivate the live vaccines (but boiling it for 15 minutes will remove most of the chlorine). In addition to being stored cool, the vaccines should also be kept away from direct sunlight. Soap must be completely rinsed off the equipment as it might inactivate the live vaccines. Also, in many countries the legislation prohibits others than veterinarians to use injectable vaccines.

Regarding the chicken house, the 1 cm gap in between the slats of the floor is likely to be too big for 14 days old chicks so they may hurt themselves. The solution could either be to reduce the spacing or to keep the chicks for a longer time in the brooder.

Integrated fish-pig farming in India

S.D. Tripathi and B.K. Sharma

Fish-pig farming material flow

The raising of pigs can fruitfully be combined with fish culture by constructing animal housing units on the pond embankment or over the pond in such a way that the wastes are directly drained into the pond. The system has obvious advantages:

- The pig dung acts as excellent pond fertilizer and raises the biological productivity of the pond and consequently increases fish production.
- Some of the fishes feed directly on the pig excrete which contains 70 percent digestible food for the fish.
- No supplementary feed is required for the fish culture, which normally accounts for 60 percent of the total input cost in conventional fish culture.
- The pond dikes provide space for erection of animal housing units.
- Pond water is used for cleaning the pigsties and for bathing the pigs.
- The system cannot be adopted in all parts of India due to religious consideration but it has special significance in certain areas as it can improve the socioeconomic status of weaker rural communities, especially the tribals who traditionally raise pigs and can take up fish-pig farming easily.

Fish-pig farming material flow

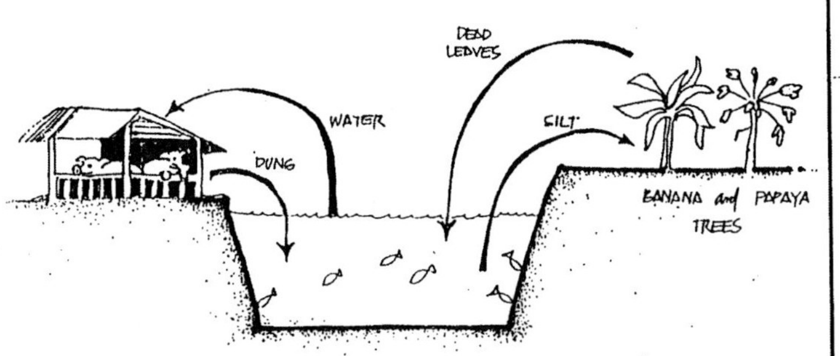

Culture practices

The ponds measuring about 1 000 m² may be located near your house, so that you can take care of the fish and pigs and can discourage poaching.

Check and repair the dikes. The pond should be deep enough so as to retain more than 1 m water depth during the dry period.

Pond preparation

Drain and dry the pond to remove all the weeds and fish fauna remaining in the pond. If it is not possible to drain the pond, all the fish can be killed by applying 15 kg of both bleaching powder and urea for a 1 000 m² pond. Alternatively, 250 kg Mahua oil cake can be applied which kills all the fishes and also acts as organic pond fertilizer.

Pigs are brought to the pond before stocking the fish, so no basal application of manure is required.

Stocking

Stock the pond with fingerlings 7 days after poisoning with bleaching powder. The recommended rate of stocking is:

CATLA

ROHU

MRIGAL

SILVER CARP

GRASS CARP

COMMON CARP

6 species		Stocking rate
Catla	-	160
Rohu	-	160
Mrigal	-	120
Silver carp	-	160
Grass carp	-	90
Common carp	-	120
Total		810

3 species		Stocking rate
Catla	-	320
Rohu	-	240
Mrigal	-	240

Alterations can be made on stocking density and species ratio, depending on the local conditions.

Grass carp should be fed regularly with aquatic or terrestrial vegetation. Liming of the pond is done at regular intervals. It helps in stabilization of organic matter. About 25 kg lime shall be required for one year.

Harvesting

Due to abundance of natural food in the fish-pig pond, the fish attains marketable size within a few months. Partial harvesting, therefore, should be done three times, depending upon the growth of fish. Final harvesting may be done after 10-12 months.

Pig raising

The number of pigs required will depend upon the pond area. The excreta of three pigs are sufficient to fertilize a pond of 1 000 m². So three pigs may be raised on a pond of 0.1 ha. As pigs attain slaughter size within 5-6 months and fish raising of Indian exotic carp is done for 10-12 months, two lots of pigs can be raised along with one lot of fish.

The pigsties are constructed on the pond embankments in such a way that the washings are drained to the pond through a delivery channel. A diversion channel is always provided to divert the excreta away from the ponds as these develop algal bloom or any other abnormality. Washings of pigsties are drained into the pond after sunrise to avoid oxygen depletion.

The pigsties can be constructed from any available cheap materials but the floor must be cemented with a slope towards the pond. Each pig is provided with a floor space of 1-1.5 m².

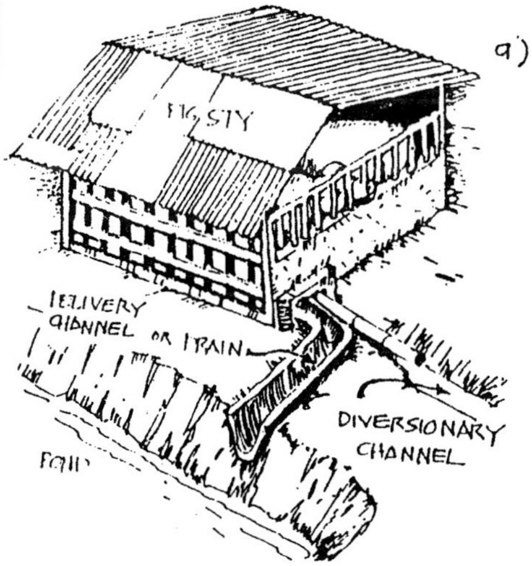

A) SAMPLE ILLUSTRATION OF PIG STY LOCATED ON A POND DIKE
B) ILLUSTRATION SHOWING SLOPING FLOOR

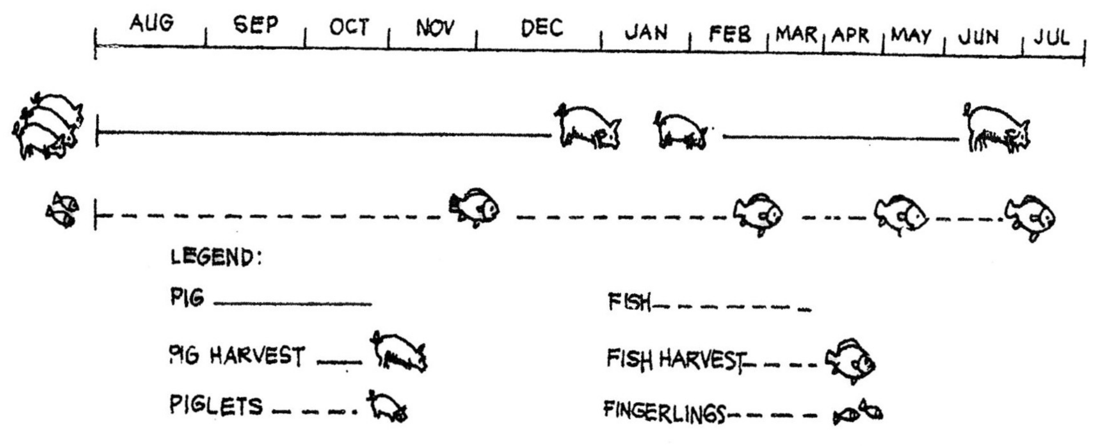

Calendar of activities for fish-pig farming

August	Pond preparation, erection of pigsties, raising of piglets
September	Stocking fingerlings, fattening and care of pigs
October	Fattening and care of pigs and fish culture
November	Fattening and care of pigs/and fish culture
December	First partial harvesting of fish
January	Harvesting of first lot of pigs
February	Fattening of second lot of pigs
March	Second partial harvesting of fish
April	Fattening of pig and fish
May	Third partial harvesting of fish
June	Preparation for final harvesting of pigs and fish
July	Final harvesting of fish and second lot of pigs

Budget (in rupee) for fish-pig farming in 0.1 ha pond

Cost	
Fish culture	
Pond preparation with 15 kg bleaching powder and 15 kg urea at Rs4.15 kg	125
Fingerlings (850 at Rs250/1 000)	213
Liming (25 kg at Rs2/kg)	50
Nets and labour for harvesting	400
Fish culture equipment	25
Pond rental	300
	1 113
Pig raising	
Pigsties	1 000
Fattening of two lots of piglets for 6 months each	
- First lot	
Weaned piglets (3 at Rs30 each)	90
Pig feed (540 kg at Rs2/kg)	1 080
- Second lot	
Medicine	40
	2 210
Total cost	3 325
Interest on the working capital at 15%	499
Total operational cost	3 824
Income	
Fish sale (600 kg at Rs20/kg)	12 000
Meat sale (300 kg live weight at Rs17/kg)	5 100
	17 100
Balance	13 276

Cash flow for integrated fish-pig farming for 0.1 ha pond

	Aug	Sep	Oct	Nov	Dec	Jan	Feb	Mar	Apr	May	Jun	Jul
Inflow	0	0	0	0	+2 000	+2 550	0	+2 000	0	+2 000	0	+8 550
Outflow	-1 402	-375	-162	-172	-252	-262	-272	-262	-262	-262	-182	-111
Netflow	-1 402	-1 777	-1 939	-2 111	-363	+1 905	+1 633	+3 371	+3 109	+4 847	4 667	+13 106

Notes:

1.Cash inflow starts in December when the partial harvesting of fish is done.

2.Harvesting of first lot of pigs increases the cash flow in January.

3.Cash inflow in March. May is due to second and third partial harvesting of fish.

Issues for further considerations

Extremely resource-poor households may find it difficult to adopt the technology, as this requires the pigs to be penned up. In small-scale rural farms, pigs are typically permitted to roam and scavenge for their feed as this avoids the investment and effort of penning and then providing feed. On the other hand, in farming systems where pigs are penned, this technology will be more applicable.

In constructing the pig pen adjacent to the pond, it should be considered that urine contains a high proportion of the waste value, and water-resistant flooring would be required, but this may be expensive or unaffordable. Plastic sheeting under wooden slatted floors has been used successfully to allow collection of animal urine for use in ponds.

Another country for which an example of this technology can be given is northern Viet Nam where the potential of backyard pig-fish rearing is even more successful. This system has been well studied and analysed.

Backyard integrated pig-fish culture in the Philippines

Frank V. Fermin

Integrated pig-fish culture is not a new concept; it has been practiced for many years in many parts of Asia. Raising pigs and fish at the same time has several advantages:

- Fish farmers can produce fish without feeding and hauling manure to fertilize the pond.
- Pig-fish culture maximizes land use by integrating two farm enterprises in the same area.
- The fishpond serves as a sanitary disposal place for animal wastes.
- Backyard integrated pig-fish culture provides additional income and a cheap source of animal protein for the family.

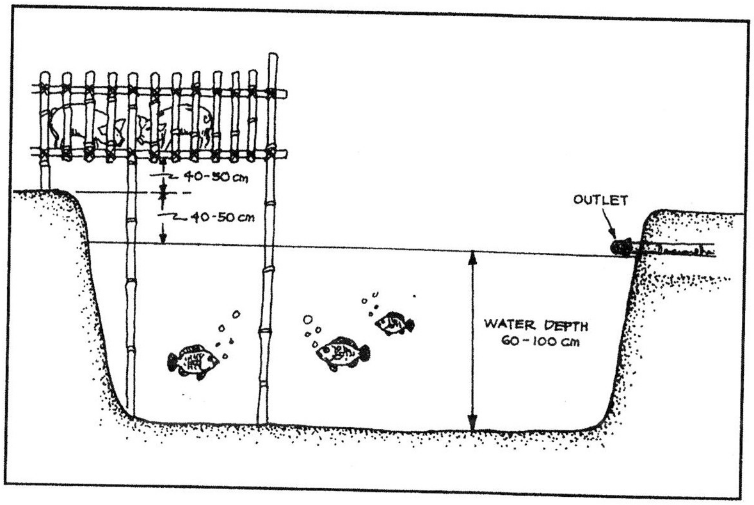

Establishing the system

1. Pond construction
   - Establish the pond near a water source. However, the site should be free from flooding. Inlet and outlet pipes should be installed and screened.
   - One pig can sufficiently fertilize a 100-150 m² pond with its manure. The water depth should be maintained at 60-100 cm. With this recommended pond area and water depth together with the right stocking density, problems of organic pollution are avoided.
   - A diversion canal can be constructed to channel excess manure into a compost pit or when manure loading needs to be stopped.

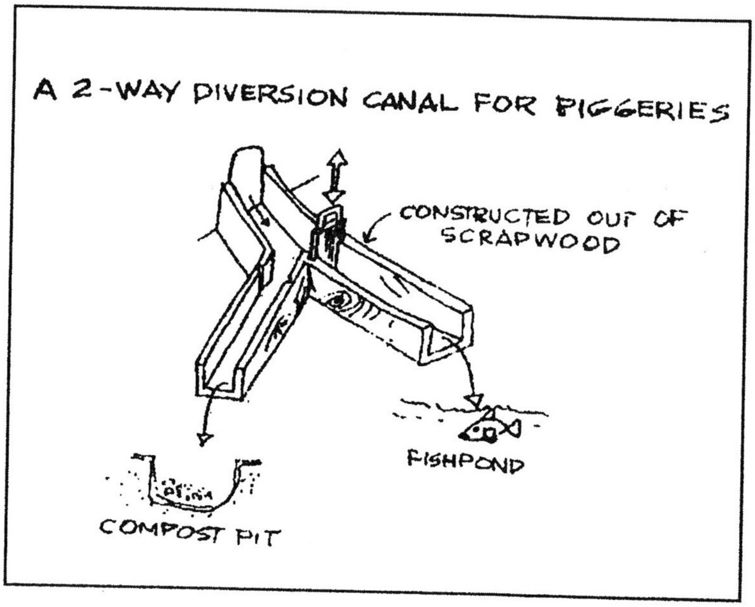

   - Nutrient-rich water from the pond can be used for vegetables grown on the pond dike or adjacent to the pond.
2. Location of the pig pen

   There are two optional designs for locating the pig pen. It can be constructed on the dikes near the fishpond. Preferably, the floor should be made of concrete (or other impermeable material to catch pig manure and urine) and should slope toward the pond. A pipe is necessary to convey the manure and urine into the pond. An alternative design is to construct the pig pen over the pond. In this case, the floor can be made of bamboo slats spaced just enough to allow manure and urine to fall directly into the pond but not too wide for the feet of the pigs to slip into (thus, causing injuries). The pen should have a floor area of 1 m x 1.5 m for each pig.

3. Stocking
   - Stock the pond (approximately 100-150 m²) with fingerlings (200 fish/100 m² once it is filled up with water. Three optional fish culture systems are suggested here, of which Polyculture 2 is based on experience in Viet Nam and Thailand. Both polycultures contain predators to control tilapia recruits (if these are mixed-sex). The recommended stocking rates are as presented in Table 1.
   - Stock the pig pen with one weanling (8-10 kg or 1.5 month old).
   - Fish and piglets can be stocked at the same time.
4. Pig feeding
   - Feed the pigs twice a day. Supplemental feeds such as kangkong *(Ipomoea aquatica)* may be given.
5. Harvesting
   - Harvest the fish after 4-5 months. Collect fingerlings (if present) for the next growing season; sell the surplus. Partial harvesting for family consumption can also be done as needed.
   - Sell the pig after 4-5 months.
   - If possible, scrape out the organic waste or mud on the pond floor and use as fertilizer for the vegetable crop.

Design 1: Pig pen on pond dike

Design 2: Pig pen over pond area

Table 1. Suggested stocking rates

Monoculture: 100% tilapia *(Oreochromis niloticus)* - 2 fish/m², 3-5 g average weight

Polyculture 1:
- 85% tilapia - 170 fingerlings, 3-5 g average weight
- 13% common carp (*Cyprinus carpio*) - 26 fingerlings
- 2% snakeheads *(Channa striata)* and catfish (*Clarias batrachus)* - 4 fingerlings, 1-2 g average weight

Polyculture 2:
- 50% *Pangasius micronemus* - 100 fingerlings, 10 g
- 30% tilapia - 60 fingerlings, 3-5 g
- 20% kissing gourami (*Helostoma temmincki)* - 40 fingerlings, 1-2 g average weight

Limitations

- High cost of inputs for pig growing (feeds and weanlings)
- Consumers may be reluctant to eat fish produced in manure-loaded ponds, creating potential marketing problems.
- Farmers want their animals close to their homes (because of theft problems) and this may not be always possible.

Possible solutions to overcome some of the limitations

1. Raise crossbred/native pigs to reduce feed cost.
2. Occasionally, fish from ponds, which were overloaded with manure, can have a "muddy" or off-flavour taste which can be removed through the following measures:
   - Stop loading manure to the pond a few days before harvesting fish.
   - Transfer harvested fish to a net enclosure installed in a clear pond at least 4-6 hours (better several days) prior to selling or eating them.

Cost and return (in Philippine peso) of the backyard integrated pig-fish culture for a 5-month period

Costs	Peso
Pig component	
Weanling	1 000.00
Commercial feeds	1 246.60
Medicines	34.00
Rice bran (P25/kg)	87.50
Labour	300.00
Pig pen maintenance	50.00
Fish component	
Pond maintenance	250.00
Fingerlings	40.00
	3 008.10
Income output	
Pig (1 head)	3 050.00
Fish (27.5 kg at P40/kg)	1 100.00
Fish fingerlings (1 100 at P.02/piece)	220.00
	4 370.00
Balance	1 369.90
Capital investments (fixed items)	
Pig pen (P500 at 6 years)	500.00
Pond construction (110-150 m²)	200.00
Bucket	80.00
	780.00

$$\text{Rate of return on investment} = \frac{1\ 369.90}{780} \times 100 = 176\%$$

Notes:
1. For P100 invested, the farmer gets P176.00
2. Entire capital cost can be recovered in one production cycle and still retain a surplus
3. 1992: US$1 = P26

Issues for further consideration

Given the better acceptance and adoption rates of this technology in other countries in Asia, there seem to be specific constraints to this in the Philippines. Pigs are resource-intensive and need a concentrate-based diet for them to grow and produce quality wastes for fishpond fertilization. Experiments have shown that *leucaena* (ipil-ipil) leaves can be toxic to pigs at relatively low levels. Growing crossbred pigs and native pigs can be fed lower-quality feeds but this does not necessarily result in overall lower feed cost per unit of pig weight produced. Growth is poorer and even if feeds are not purchased, effort and other resources are needed in their use. Pig production is often affected by marketing risks and problems, which should be considered by new entrants.

The scraping out of pond mud for crop and vegetable fertilization is labour-intensive and also requires a drained pond, which is not possible in many locations where rural fishponds have been sited.

RICE-FISH SYSTEMS

Low-input rice-fish farming system in irrigated areas in Malaysia

Ahyaudin Ali

Large-scale transect of the north Kerian irrigated area of Perak, Malaysia

SOIL	CLAY LOAM	SWAMPY	CLAY	LOAM	CLAY LOAM →					SANDY LOAM
LAND USE	RICEFIELD	ROAD/ HOUSES	CANAL	ROAD/ HOUSES	RICEFIELD	SUMP POND	RICEFIELD	ROAD/ HOUSES	CANAL	ROAD / HOUSES
LIVESTOCK		POULTRY GOAT CATTLE		POULTRY GOAT CATTLE				POULTRY GOAT CATTLE		POULTRY GOAT CATTLE
FISH			MUDFISH CATFISH GORAMIES		SNAKEHEAD CATFISH GORAMIES	SNAKEHEAD CATFISH GORAMIES	SNAKEHEAD CATFISH GORAMIES		SNAKEHEAD CATFISH GORAMIES	
TREES AND CROPS	RICE	COCONUT PAPAYA MANGO BAMBOO PINANG BANANA LEMON GRASS LIME		COCONUT PAPAYA MANGO BAMBOO PINANG BANANA LEMON GRASS LIME			COCONUT PAPAYA BANANA CASSAVA SQUASH MANGO	COCONUT PAPAYA MANGO BAMBOO PINANG BANANA LIME		COCONUT PAPAYA MANGO BAMBOO PINANG BANANA LIME

The rice-fish farming system is an old tradition practiced extensively in the North Kerian area of Perak, Peninsular Malaysia. The area is an alluvial coastal floodplain and receives irrigated water from the Tasik Merah reservoir. The soil is primarily clay with some acidity problems. Rice is double-cropped and short-season; high-yielding rice varieties are used.

There are approximately 352 000 ha of ricefields in Peninsular Malaysia, out of which 120 000 ha (34 percent) have sufficient water depth (15-16 cm) for rice-fish system.

Description

- Two systems are distinguished: the traditional (System 1) and the improved (System 2). Both systems, which are essentially trapping and ongrowing of wild fish, require little labour and material inputs. Fish enter with water from the irrigation canals, ditches and ricefields and are trapped early in the rice-growing season, grown together with rice, and later harvested at the end of the season. In the improved system, more trenches are added, sump ponds are deepened, manured and limed, and dikes are improved, which all require additional investment.
- Ricefield sizes vary from 0.81 to 1.42 ha. A sump pond, which can also be an unused well or burrow pit, ranging in size from 6.5 to 8.0 m diameter, is located at the lowest part of the field. Sump pond, which is cooler and has higher dissolved oxygen content, provides shelter for fish during periods of low water level. Perimeter trenches (0.25 m wide and 0.1 m deep) may be dug around the field to enable fish to move to and from the sump pond. These open trenches also act as feeding areas for fish to feed upon zooplankton especially during the fry and fingerling stages when zooplankton is important for early growth. Mud obtained from digging the trenches is used to strengthen the dikes (0.3 m high) around the field.
- No supplementary feeding is provided. Fish obtain food from natural sources in the ricefields. The system fertility

Transect of the north Kerian irrigated area of Perak, Malaysia, with all resource types of traditional and improved systems

LAND USE	IRRIGATION DRAINAGE CANAL	ROAD/ DIKE	DITCH	HOMESTEAD	RICEFIELD	DIKE	SUMP POND	DIKE
LIVESTOCK				GOATS POULTRY CATTLE				
FISH	SNAKEHEAD CATFISH GORAMIES		SNAKEHEAD CATFISH GORAMIES		SNAKEHEAD CATFISH GORAMIES		SNAKEHEAD CATFISH GORAMIES	
TREES AND CROPS				BANANA, MANGO, COCONUT, PAPAYA, BAMBOO, LIME, LEMON GRASS		BANANA CASSAVA SQUASH OKRA STRING BEANS PAPAYA CHILI		COCONUT BANANA MANGO
SYSTEM IMPROVEMENT	GENE POOL FOR FISH		GENE POOL	BIO-INTENSIVE GARDENING	MAINTAIN DIKES, TRENCHES, REDUCE PESTICIDE AND HERBICIDE USE	BIO-INTENSIVE GARDENING	GENE POOL STOCK NEW SPECIES, SUPPLEMENTARY FEEDING, MAINTAIN POND	BIO-INTENSIVE GARDENING WITH ADDED COSTS
SYSTEM LIMITATION	COMMON PROPERTY		COMMON PROPERTY	LABOR SHORTAGE	LIMITED LAND FOR PHYSICAL IMPROVEMENT UNCONTROLLED FLOODING EXCESSIVE USE OF PESTICIDE OVERHARVESTING		POACHING, LOW PRODUCTIVITY, NO FISH STOCKING, UNCONTROLLED FLOODING, EXCESSIVE USE OF PESTICIDE OVERHARVESTING	

Farm layout of traditional system

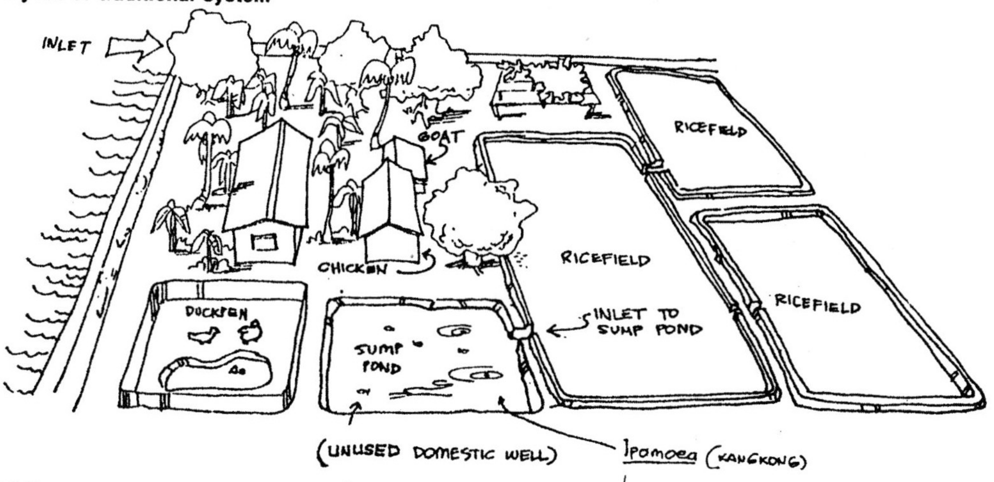

Material flow on a traditional farm system

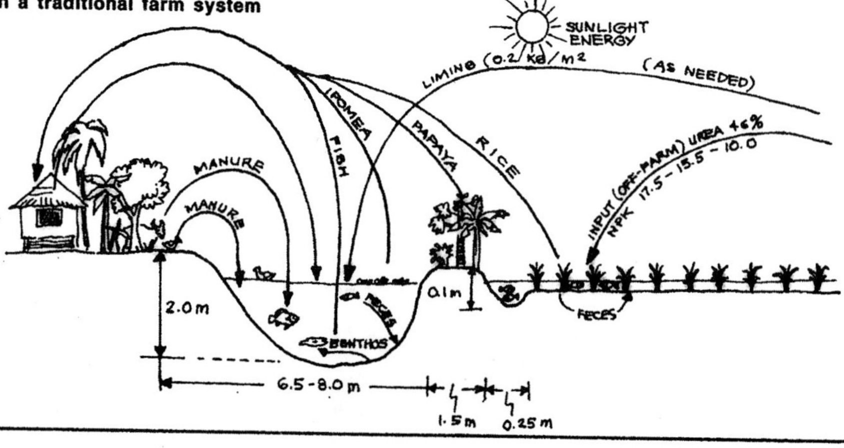

Farm layout of improved system

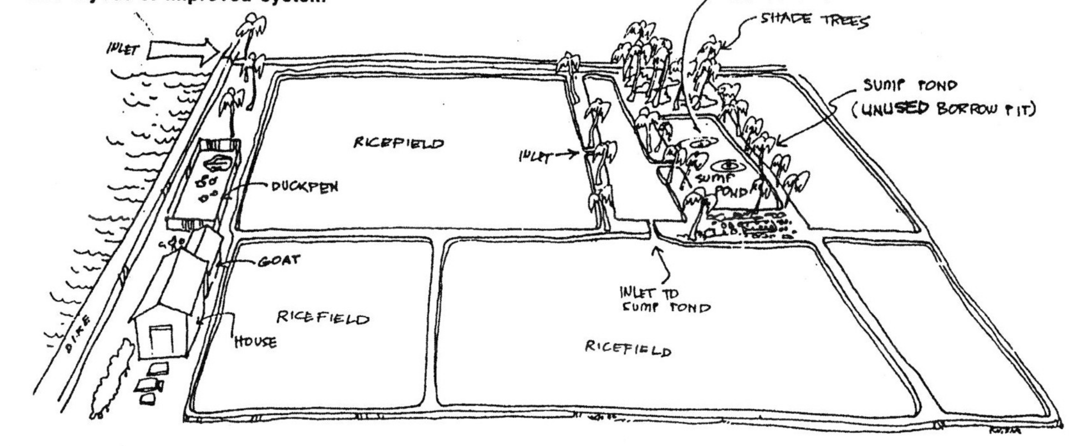

Material flow of an improved farm system

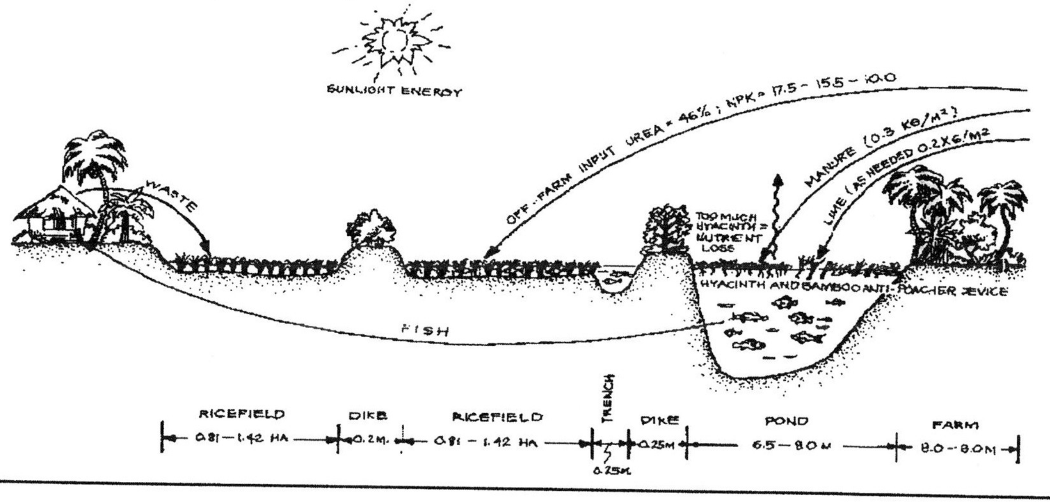

depends on rice fertilization which is applied twice during the growing season. Urea (46 percent N) and NPK (17.5-15.5-10.0) fertilizers are used at the rates of 56 and 112 kg/ha, respectively. To further increase productivity and food availability, liming of the sump pond (if required) and manuring (if available) should be done.

Species used

Local species

Local species grow well in ricefields. They have adapted to shallow water, high turbidity and temperature, and low dissolved oxygen conditions of the fields.

- The snakeskin gourami (*Trichogaster pectoralis*) is numerically the most important. This species and the three-spot gourami (*T. trichopterus*) are herbivore/planktivore and occupy the lower rung in the food chain.
- Climbing perch (*Anabas testudineus*) is an insectivore.
- The catfish (*Clarias macrocephalus*), an omnivore, and the mudfish or snakehead (*Channa striata*), a carnivore, are also important species.

Newly considered species

- Tilapia (*Oreochromis* spp.), a herbivore/planktivore/insectivore, is ecologically suitable and economically important.
- Another species to be considered is the freshwater prawn (*Macrobrachium rosenbergii*).

Yields

- Fish sales provide an important supplementary income especially to tenant farmers. Contributions from fish are 6.8 and 9.0 percent for owner and tenant farmers, respectively. Because little inputs are involved, the yields contribute significantly to farmers' seasonal income.
- Fish are sold to dealers who provide pump, nets and other accessories needed to harvest fish. Small fish are left behind as stock for the next growing season. Marketable-size fish are: snakeskin gourami, 14 cm; catfish, 20 cm; and snakehead, 25 cm.

Seasonal calendar

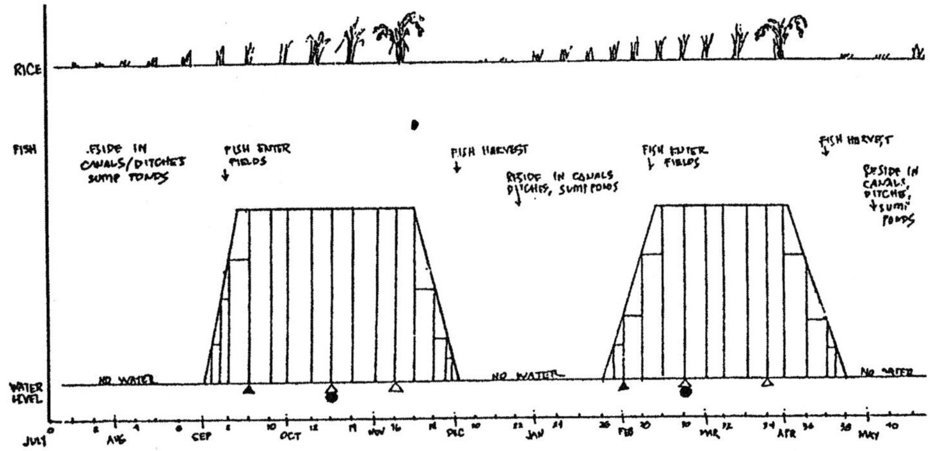

▲ Field preparation (herbicides application). Caution: use properly and only when necessary
△ Fertilization (urea-56 kg/ha; NPK-1.2 kg/ha)
● Pesticide application (Carbofuran-5.6 kg/ha). Caution: use properly and only when necessary

Advantages

- Provides additional food and income
- No additional expenses except when system is modified such as building trenches, strengthening dikes, etc.
- No major changes in normal farm practices: in the improved system (System 2), modifications to improve yields are adapted to traditional practices by farmers if affordable. The traditional system (System 1) is sustainable for limited labour/ older couple situations.
- Optimises disused and under-utilized existing resources
- Maintains gene pool for locally valuable species

Constraints

- Short growing season due to double dropping of rice
- Improper and excessive pesticide and herbicide use
- Uncontrolled flooding may result to fish loss
- Improper management/lack of human resources
- Low productivity and low-carrying capacity when no regular supplementary feed is provided

Calendar of activities

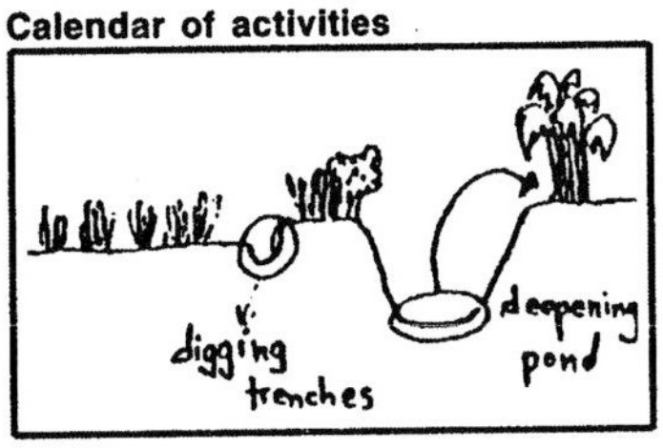

- **During fallow season sump pond should be deepened once every five years to about 1.5-2.0 m. Perimeter trenches (0.25 m wide and 0.1 m deep) should be dug. Dike should be raised to about 0.4 m. Liming if needed should be done and manure if available should be applied.**

② 0 days

- **Once water is available, field preparation should be done by cutting, raking and removing dead weeds. In about 7-10 days, all exits should be blocked to prevent fish from escaping.**

③ 20 days

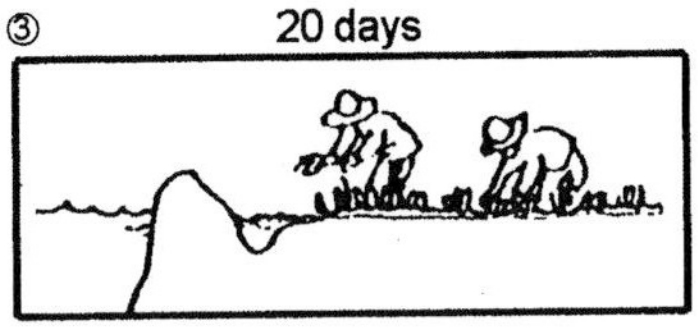

- **Transplanting of rice seedlings. Weeding should be done especially in the trenches to provide feeding area (on plankton) for young fish.**

④ 45-50 days

- **First fertilization. Add 5.6 kg/ha Furadan (Carbofuran) mixed with urea (56 kg/ha) and NPK (112 kg/ha).**
- **Second fertilization after 60-65 days. Same as above.**

⑤ 65-150 days

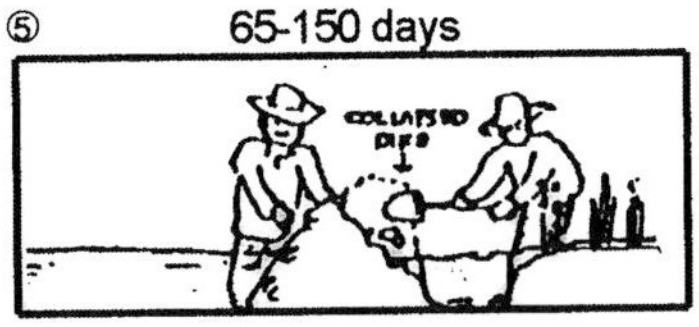

- **Check the ricefield as well as the fish. Stop all leakages to prevent fish from escaping.**

⑥ >150 days

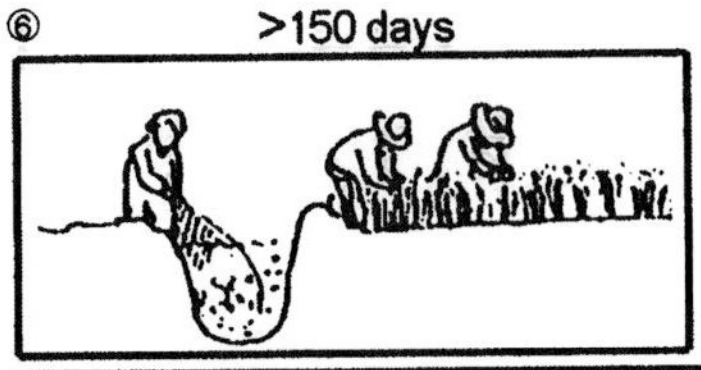

- **When rice is about ready to be harvested, drain the pond to harvest the fish. Take only marketable-size fish (size depends on market demand). Leave smaller fish as stock for next season.**

Estimated range of potential fish yields obtained from traditional and improved rice-fish systems

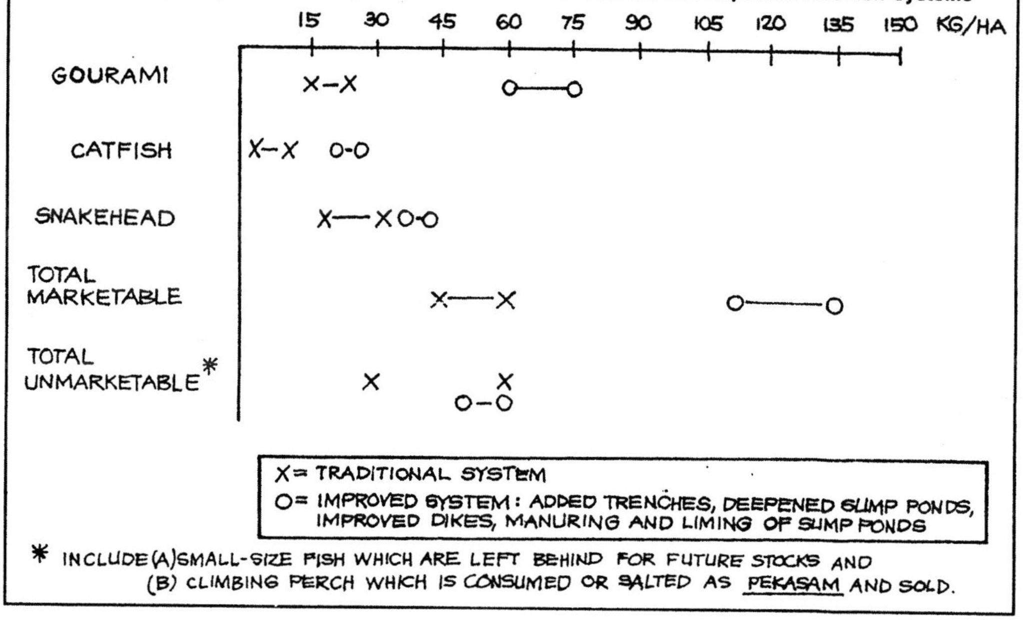

Estimated simplified budget (in US$) for the fish component of the two rice-fish systems

	Season 1		Season 2	
	System 1	System 2	System 1	System 2
Yields (ranges)				
Cash from fish sales	26-35	47-59	26-35	47-59
Cash from *pekasam*	8-12	10-12	8-12	10-12
Sales (4-6 kg x 5-6 kg/season, US$2/kg)				
Noncash (domestic consumption - mostly snakeskin gourami, 10-20 kg/ season)	1-3	1-3	1-3	1-3
Total (a)	35-50	58-74	35-50	58-74
Costs				
Cash	35-50	58-74		
Fixing dikes, trenches and sump				
Pond and other management steps (4.5 day x US$4)				
Materials		18		
* Manure (12.5 kg x US$0.800)		10		10
* Lime (8.5 kg x US$0.80)		7		7
Total (b)		35		17
Net income (a-b)	35-50	23-39	35-50	41-57

Notes:
1. Notice how in the first season, System 1 has no initial costs and thus a higher net income than 2.
2. However, in the second season, the costs for System 2 go down and farmer gets a higher return compared to 1.

- Distance of sump pond from house
- Conflicting government programs in the form of subsidy for rice

Issues for further consideration

The technology requires relatively low amounts of material and labour inputs and hence is useful for low-income households.

In designing new applications of the system elsewhere, the following information may be necessary: (a) what are the roles of the newly introduced fish species; (b) elsewhere, tilapia recruits are a valuable source of feed for the snakehead leading to overall yield increases when these recruits are present; (c) does the improved system increase the inventory of fish stored between crops; and (d) which rice practices are being used: broadcasting transplanting or direct seeding?

Rice-fish systems in Indonesia

Catalino dela Cruz

Rice-fish farming has a long history in Indonesia. In general, farmers have developed the systems that are now used. The widely practiced rice-fish systems in irrigated areas of West Java are *minapadi, penyelang* and fish *palawija.* A special system called *sawah tambak* also exists in the coastal areas of East Java.

Fish produced from ricefields are mostly seedfish for restocking in growout systems, such as floating net and bamboo cages, running water (concrete tanks) and irrigation canal systems.

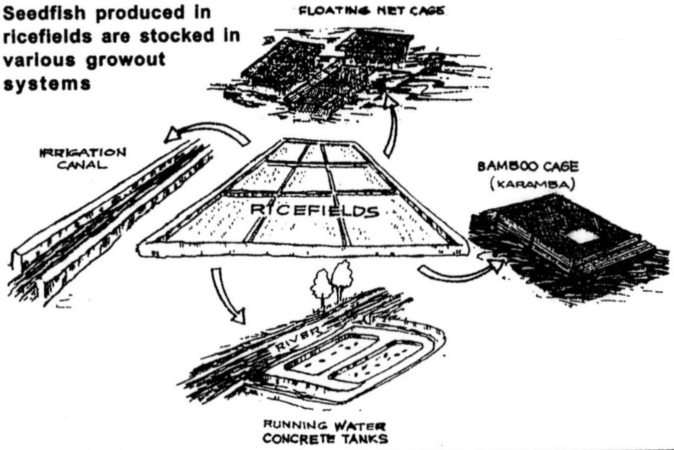

Seedfish produced in ricefields are stocked in various growout systems

Minapadi system

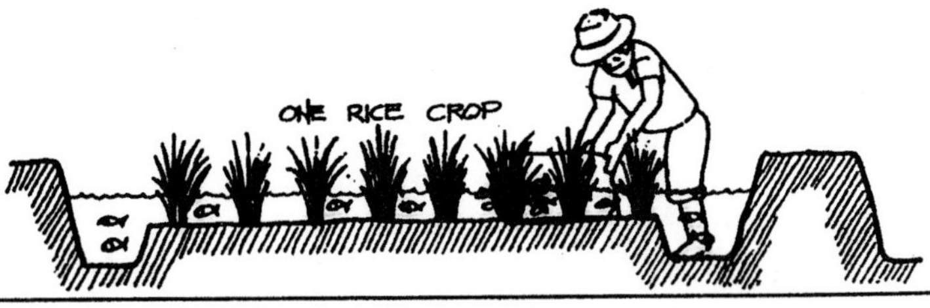

Minapadi system

In this system, rice and fish are concurrently raised in the same area. A trench refuge (0.5 m wide and 0.3-0.4 m deep) is used. The *payaman* method is another kind of *minapadi* system. The difference is that the rice-fish field is connected to a pond refuge instead of a trench.

Rice varieties that are proven to give high yields with fish during the wet season such as IR 64 and during dry season such as *Ciliwung* are planted. Planting distance in a thoroughly prepared plot is 20 x 20 cm, 22 x 22 cm or 25 x 25 cm. In West Java, fertilizers used (and their rates of application in kg/ha) are: urea, 200; triplesuperphosphate,100; potassium chloride, 100; and ammonium sulphate, 50. Water level is kept low during the tillering stage of rice. It is gradually raised to 10-15 cm throughout the rice growth period.

Common carp weighing 15-25 g are stocked in the ricefield at 2 500-3 000/ha 7-10 days after rice transplanting. A center or cross-trench occupies about 2 percent of the total ricefield area. Harvesting is done by draining the field slowly after a culture period of 40-60 days. Within this period, the fish attain 50-100 g, the size desired for stocking cages and running water culture systems.

Penyelang system

This is the culture of fish in between the first and second rice crops. The fish culture period is shorter than in *palawija* system. A portion of the ricefield with rice stubbles is immediately stocked with common carp, while preparing the remaining portion to the dry season rice crop.

Stocking size varies: 5-8 or 8-12 cm or 15-25 g, depending on availability. Stocking rate is 2 000-4 000/ha. Water depth is 10-20 cm. Fish are harvested after 30-40 days. This short period may not produce the desired size for growout in cages and running water systems, especially if stocked small. However, growout operators also buy small fish seeds if supply is scarce. The unsold small fish are restocked in the following dry season rice crop cycle.

Penyelang system

Palawija ikan system

Immediately after the harvest of the dry season rice crop, dikes are raised by using a hoe, to contain water depth of 30-40 cm. The stocking size and rate vary. In West Java, common fish carp of size 3-5 or 5-8 cm are stocked at 5 000/ha without feeding. In North Sumatra, consumption size is produced in the *palawija* system. The usual common carp sizes stocked are 30-50 g or 50-100 g at the rate of 1 000-1 500 (no feeding); and 1 500-3 000 (with supplemental feeding). Supplemental feeds are rice bran, chopped cassava, corn kernel soaked in water, poultry feed, kitchen refuse and others. Harvesting the fish is done by draining the field.

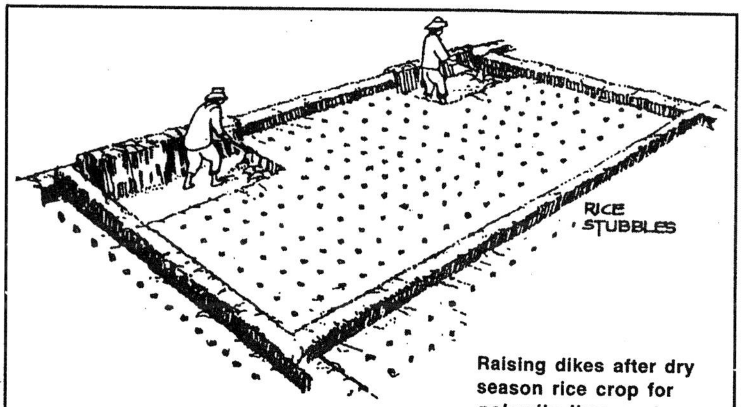

Raising dikes after dry season rice crop for *palawija ikan* system

Fish stocking and production data

System	Stocking size	Stocking rate (n/ha)	Production (kg/ha)	Culture period (days)
Minapadi	15-25 g	2 500-3 000	100-200	60
Penyelang	15-25 g	2 500-3 000	70-100	30-40
Palawija	8-5 cm	5 000	200-300	60
	30-50 g 50-100 g	1 000- 3 000	300-800	60-70

Minapadi (rice+fish) - *penyelang* (fish only) - *minapadi* (rice+fish) - *palawija* (fish only)

Rice - *penyelang* - rice - *palawija*

Rice - rice - *palawija*

(Rice+fish-duck) - (fish-duck) - (rice+fish-duck) - (fish-duck)

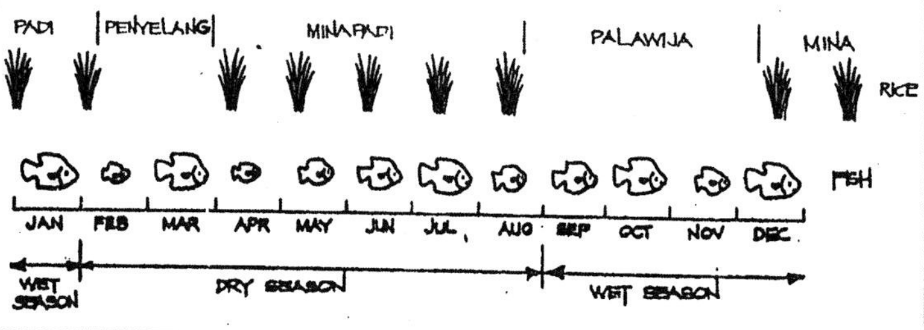

Cropping patterns

The above systems are combined into sequential cropping patterns in a year such as:

In the last pattern, ducks are allowed to roam in the ricefield 25-30 days after transplanting the rice. Ducks have potential for controlling golden snail *(Pomacea spp.)* infestation on rice at a density of 25 ducks/ha. The ducks have a small refuge pond where they are kept when necessary.

The addition of ducks in the last pattern made it the most profitable pattern. The year-round supply of eggs provides monthly income to a farmer. In the absence of ducks, the *minapadi-penyelang-minapadi-palawija* pattern is the most profitable.

Issues for further consideration

For the system described, potential users may need to obtain local information about the relative importance of different markets for fish seed, and the quantity and schedules these are needed.

The practice is concentrated in West Java. Wealth ranking of the types of farmers involved (previously and present-day), as well as the inputs provided by and the access to benefits by different household members, would provide useful information for further applications. Also of interest are the local variations in the relative value of fish and rice, in comparison to the situation in West Java by main system type.

Economics of the systems should be evaluated based on local trials, considering local species needs, costs and returns. Generally the system may find successful application where there is adequate demand for fingerlings.

Additionally the system can bring with it employment opportunities for women, particularly in fish seed rearing, as it ensures quick return on their labour, cash and material investment.

Sawah tambak rice-fish system in Indonesia

Catalino dela Cruz

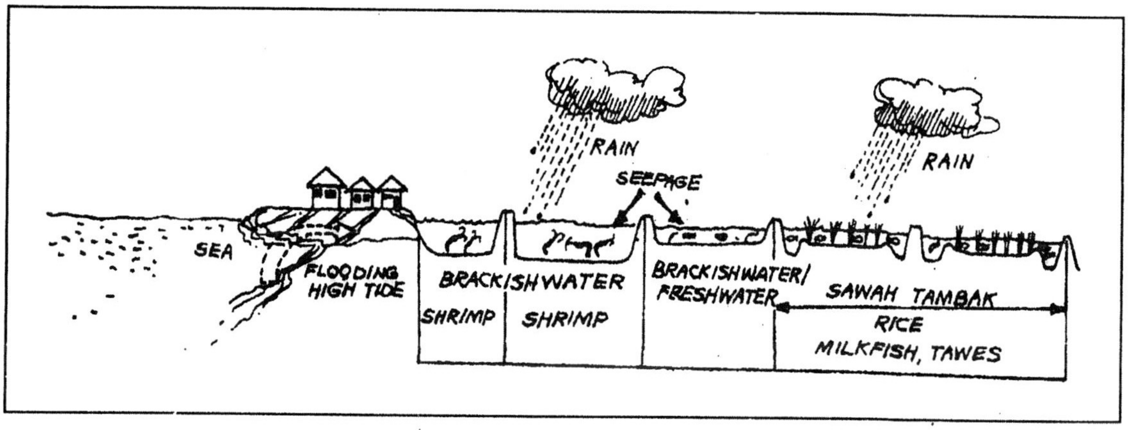

Literally, *sawah tambak* means ricefield pond (brackishwater). However, this term refers here to the 12 152 ha rice-fish farm area in East Java that involves 15 000 households. Depending on the depth of floodwater in each area, and fish or rice culture intensity, the *sawah tambak* rice-fish systems can be classified as follows:

1. Concurrent rice-fish system during wet season: appropriate in areas where inundation and the risk against submergence of rice is low. On the other hand, water is not sufficient to support a dry season rice crop.
2. Concurrent rice-fish (wet season) followed by dry season rice: done in areas where standing water is not so deep and water is sufficient to support dry season rice crop.
3. Fish culture (no rice) in wet season followed by dry season rice: appropriate in areas where flooding is deep.
4. Fish culture throughout: done in areas where farmers prefer to raise fish instead of rice in the entire flood season.

Fish species grown are a combination of milkfish *(Chanos chanos)* and tawes or silver barb *(Barbodes gonionotus)*. Common carp is also added if available.

Operation of the *sawah tambak* system

Field components

1. Peripheral dike. This is built by excavating the inner peripheral canal of the field. Base width: 4-5 m; top width: 2-2.2 m; height: 1.4-1.8 m.
2. Peripheral canal/trench. This serves as a fish refuge, nursery, holding/transition place, catching canal and source of water for dry season rice. Bottom width: 2-4 m; top width: 2.8-3.2 m; depth: 0.3-0.7 m.
3. Ricefield area. The area used for planting rice is surrounded by a temporary bund 0.5 m high. This retains the water required by rice for its growth. The bund is also needed especially in concurrent rice-fish system.

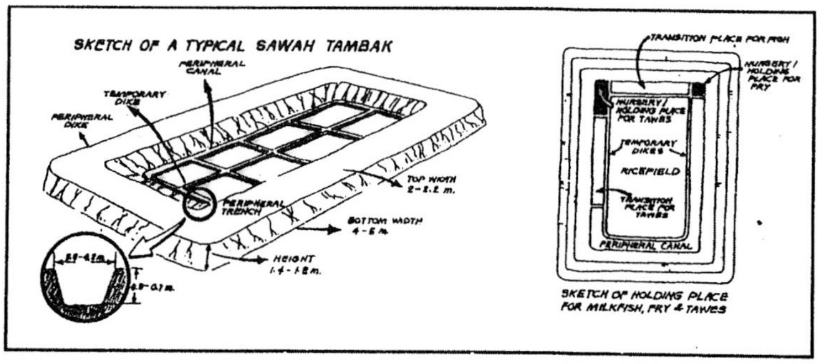

Water supply

Water comes from rainfall or seepage. Thus, there is no need to provide water inlet or outlet gates. When it is necessary to reduce or add water, pumping or bailing it out by traditional method is used.

Prevention of fish escape during floods

Farmers have ready grasses, plant leaves and similar materials to spread on top of dikes when floods overtop dikes.

Preparation of ricefield area

The ricefield enclosed by the dikes is prepared just like an ordinary one. Land preparation begins in September just before the onset of the rainy season, either by dry or wet method.

Nursery/holding and transition areas

These are constructed in the peripheral canal. The nursery is 10 m long, 5 m wide and 0.75 m deep. Water filling from outside is done through pumping or by traditional bailing method. Fry stocking is done 2-3 days after water filling.

Oftentimes, prior to stocking fish in the entire *sawah tambak*, the milkfish and tawes fry are cultured separately in a nursery/holding corner in the peripheral canal. The milkfish (stocking rate: 500 m^2) are raised here up to 45-60 days. The tawes (220 m^2) are kept at the holding place (with about 50 m^2 water depth) for one month before releasing them into the field.

Fertilization

Organic (compost, animal manure, green aquatic plants, etc.) and inorganic (urea and triplesuperphosphate) fertilizers are applied. The application rates (in t/ha/year) are:

Rice hay	10-15
Plant leaves	1-4
Green aquatic plants	2-5

Urea is applied at the rate of 100-150 kg/ha/year and triplesuperphosphate is at 300-450 kg/ha/year.

The total amount of urea and triplesuperphosphate is each divided into three equal parts and applied thrice. As an example, the first application is a mixture of 25-50 kg/ha urea and 100-150 kg/ha triplesuperphosphate.

Culture management and harvesting

Stocking sizes and rates per hectare are as follows:

Milkfish	Fry	11 000	
	5-7 cm	5 500	8 250
Tawes	Fry	22 000	
	5-7 cm	5 500	11 000

Culture period is 4-7 months, depending on the available standing water. In areas with deep water, culture period extends to one year. Stocking of fish can be done more than once. Harvesting is done twice or thrice. With no feeding, yield is about 2 000-3 500 kg/ha.

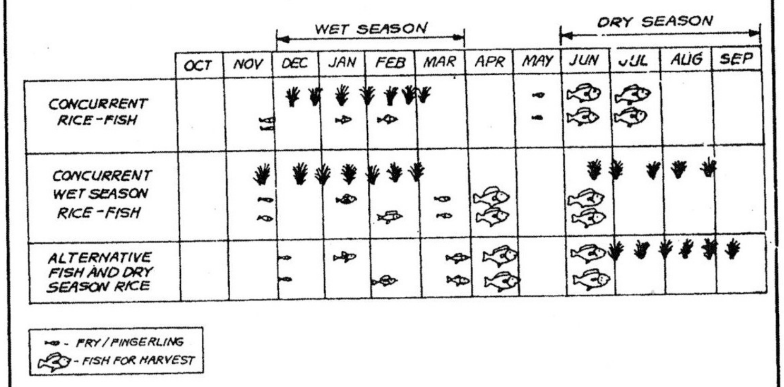

Issues for further consideration

In considering if the method is useful to adopt, other occupations of the farmers, aside from rice-fish farming, will be essential information. How important is this part of their livelihood system? This can be accessed through comparative economic resource analysis. Farmers will want to know what benefits the addition of the fish component brings, in comparison to rice culture alone. Of further interest is the access to benefits from the system within household.

The risk of floods and the possible escapes of fish should be assessed. How effective and costly are simple methods against this risk?

In rearing fish during the initial period of 6-8 weeks, more detailed information will be necessary to ensure success, also in respect to manuring and feeding schedules.

The salinity ranges in the targeted area will have influence on the choice of species and management decisions.

Rice-fish systems in China

Yixian Guo

Rice-fish farming calendars (Hubei, China)

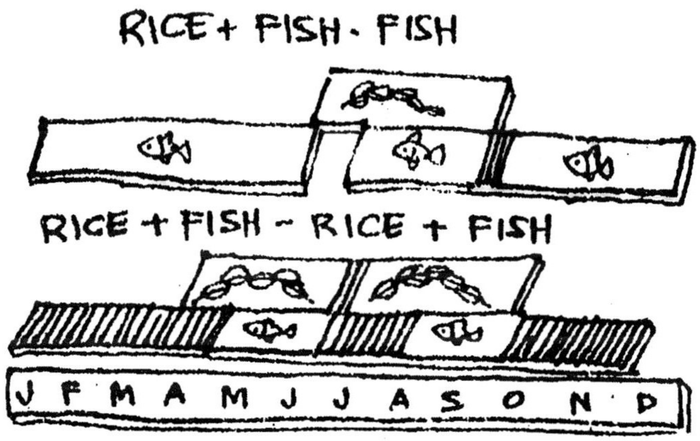

Rice-fish farming, an age-old practice in China, can be traced back more than 1 700 years, although recently it has been largely ignored. Integrated rice-fish farming in China is generally characterized by four basic components: (1) extensive use of land; (2) low input; (3) low yield; and (4) household consumption of rice-fish farming.

After the founding of the People's Republic of China, the government organized farmers and encouraged them to develop integrated rice-fish farming systems. As a result, hectarage under rice-fish farming reached 700 000 ha in 1959, but sharply declined in the 1960s and 1970s due to wide use of pesticides, reformation of cropping systems and unfavourable national economic policy during the "cultural revolution" period (1966-1976). During this period, acreage of rice-fish farming dropped from 40 000 ha to 320 ha in Guangdong province and a similar drop from 230 000 ha to 5 300 ha was documented in Hunan province. However, during the recent "reformation" and "opening" period, the government is again encouraging the adoption of rice-fish farmers.

With farmer initiative and assistance from the government, the adoption of rice-fish farming is rapidly expanding. It has traveled from Guangdong province in the south to Hei-Long-Jiang province in the north and has reached historical proportions with more than 1 million ha in 1986. Sichuan, Hunan, Guizihou and Fujian are the top four provinces in China.

Rice-fish systems are principally found in the areas of the Yangtze River basin and other parts of southern China, although some rice-fish can be seen in northern provinces. The traditional rice-fish systems presented here are found in both irrigated and rainfed areas. The improved designs are principally found in irrigated conditions. Most rice-fish farmers in China are "cooperative farmers" with small landholdings of 1 500 m^2 or less. Normal fishpond size is usually 1 000 m^2.

Major component technologies of rice-fish systems in China are presented here.

1. Appropriate construction of paddy field
   - The traditional paddy field layout has no trench or pond in the field and the water storage capacity is limited. Fish growth is more directly affected by rice crop management, and the result is a low and unstable yield.

Here are some trench-pond integrated rice-fish designs.

Design A: trench-pit design

This is an improved design with a small, shallow pit (1-2 m^2) in the center of the field.

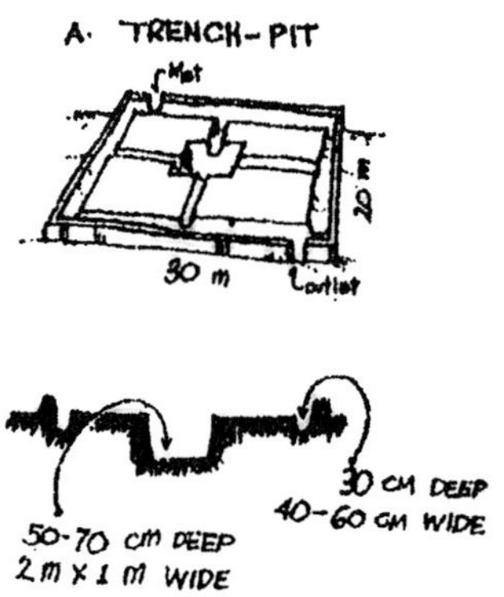

Crossing trenches are dug to connect the pit to all side trenches. Increased water storage capacity offers a better refuge for the fish. This design raises rice yield by 10 percent and 1-2 times as many fish can be raised as compared with the traditional design.

- Design B: trench-pond design

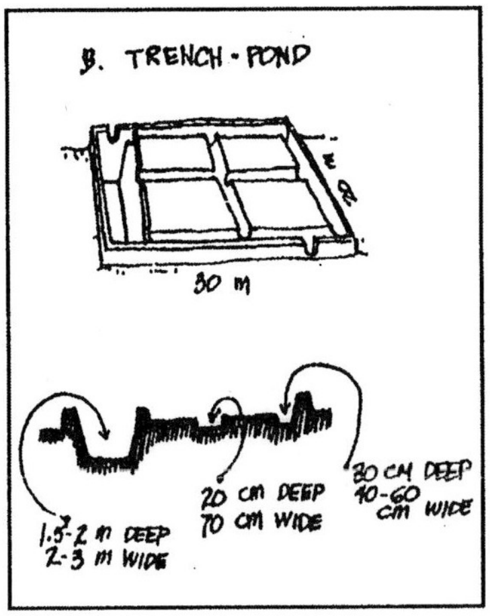

This design is a further improvement with a larger, deeper pond at one end of the field. Crossing trenches are also dug to connect the pond to all sides. This design significantly increases the water storage capacity and provides a better environment for the fish. It raises and stabilizes the yield of both rice and fish.

- Ridged-field rice-azolla-fish model

Ridged-field rice-azolla-fish model

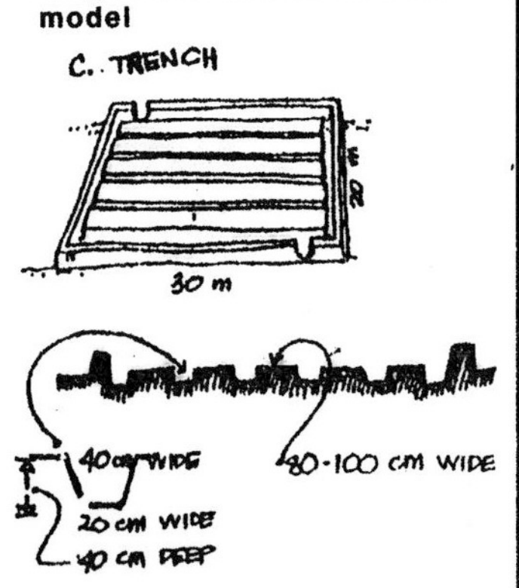

This design was originally developed for swampy areas with the objectives of improving soil properties and increasing rice yield. Later it was stepwise, integrated with azolla and fish. Rice is planted on the ridge, azolla as a feed for fish as well as a biofertilizer, and green manure and fish are stocked in the trenches.

Production data of rice-fish-azolla system (irrigated lowland)

Rice (total for two crops/year)	Fish	Azolla (fresh weight)
862 kg/mu	50.21 kg/mu	2 010 kg/mu
12 916 kg/ha	753 kg/ha	30 150 kg/ha

1 Chinese mu = 0.67 ha

Azolla is a small aquatic fern (usually 1-5 cm large) which can grow on saturated or moist soils. It is capable of doubling its weight in 3-5 days. Azolla has symbiontic bacteria which fix atmospheric nitrogen and can fix 3-7 kg N/ha daily. It contains 4 percent nitrogen on a dry-weight basis and is an excellent source of nitrogen fertilizer.

2. Basal fertilizer application

Physical injury to fish caused by inorganic fertilizers used in rice production can be a constraint to rice-fish systems. Necessary measures should be taken to minimize fish injuries. One such measure is to increase the amount of basal fertilizer application during the land preparation stage to approximately 80 percent of the total nitrogen and 100 percent of the total phosphorous requirement.

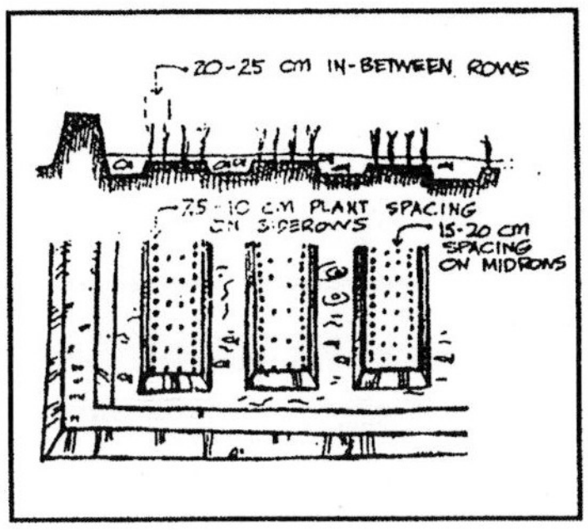

3. Transplanting

Reduced rice plant population within the ridged-field paddy design caused by the construction of trenches and refuge ponds is one farmer constraint to the practice of rice-fish systems. Farmers can lose as much as 10 percent of their paddy when constructing trenches and refuge ponds for rice-fish systems. To minimize reduced plant population (and potentially decreasing crop yield), plant spacing can be intensified by lessening the recommended distances between hills while maintaining the row spacing (20-25 cm). Normal hill spacing of 15-20 cm between plants can be cut in half, thus doubling the plant population in the side-rows of the trench.

4. Fish stocking considerations in rice-fish systems

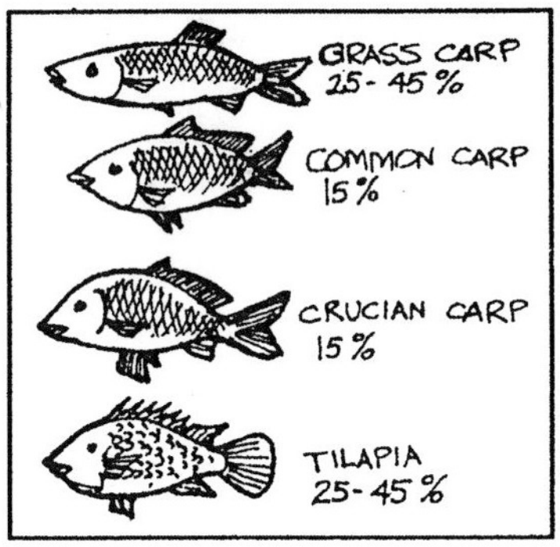

- Grass carp, common carp, Nile tilapia and crucian carp are four predominant species for polyculture rice-fish farming in China. One such system involves four species: grass carp (*Ctenopharyngodon idellus*), Tilapia sp., common carp (*Cyprinus carpio*) and crucian carp (*Carassius auratus*), with the first two species making up the majority. The recommended mix is a 25-45 percent composition of both grass carp and tilapia (a total combined composition of 70 percent) with a 15 percent mix of both common and crucian carp (remaining 30 percent) at stocking rate of 2-3 m². This mix of species can give an optimal yield of both rice and fish.

- When stocking fingerlings in the pond or paddy, a large difference in water temperature between the container used to transport the fish and the paddy can lead to fish loss and a poor survival rate. Therefore, it is recommended to mix water from the paddy with water in the container to slowly regulate the temperature differences and allow the fish to adjust to the water temperature field.

5. Top-dressing fertilizer application

Top-dressing fertilizer is applied at the panicle differentiation stage (about 28-30 days to heading). A shallow layer of standing water in paddy is necessary for fertilizer application, but can increase the possibilities of injuring the fish. However, two alternatives exist which can help minimize these limitations.

- Slowly drain the water from the paddy allowing the fingerlings to return to the refuge trench/pond. As the water in the ridges almost dried up, the top-dressing fertilization (broadcast) can be applied, thus avoiding injury to the fish as well as achieving fertilizer efficiency. Two to four days after the fertilizer is applied, the field is again flooded.
- Fertilizer application can be done by deep placement by hand or by using machinery. The fertilizer should be applied at a depth of 8-10 cm. Fertilizer efficiency and reduced risk to fish health are also attained. (Note: The paddy should also be drained for application of the deep-placement method).

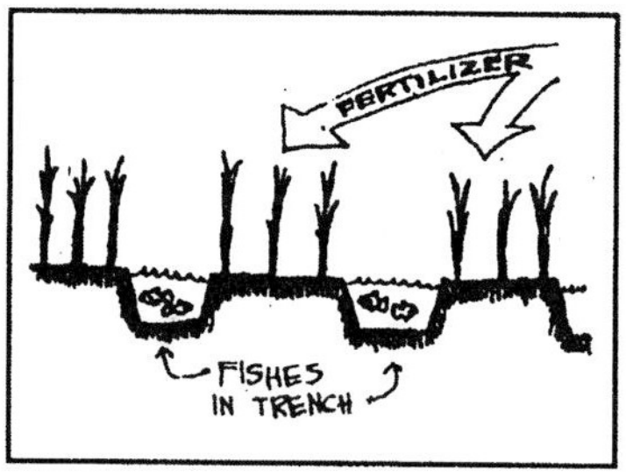

6. Pest management

- Many weed varieties found in the rice paddy are good feeds for grass carp. Stocking fry at 2-3 pieces/m² (2-3 cm long) and fingerlings at 2-3 pieces/10 m² or (8-10 cm long) one month after transplanting can help control weeds, thus, reducing the need for other weed-control techniques. As the fingerlings grow, daily supplemental feeding with green grasses is necessary to avoid damage to the young rice plants by the fish. Grass is usually fed to the fish in the pond to avoid damage to the rice plants; while rice bran and other supplemental feeds can be directly fed to the fish in the paddy area.

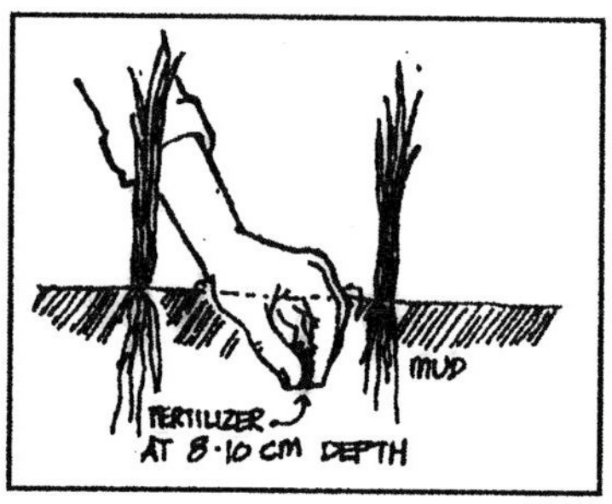

- Fish eat insects, such as stemborer and leaffolder which move through the water among the rice plants and hoppers which catch them as they float on the water from the rice plant. Fish reduce need for pesticide. They also eat the pathogens (such as sheath blight disease) floating in water and on the bottom as well as disease-infected leaves. This not only reduces the pathogens but also improves plant's health. The use of fungicides can also be reduced. If pesticides must be applied, certain precautions should be taken. Traditional fields should be flooded with more water.

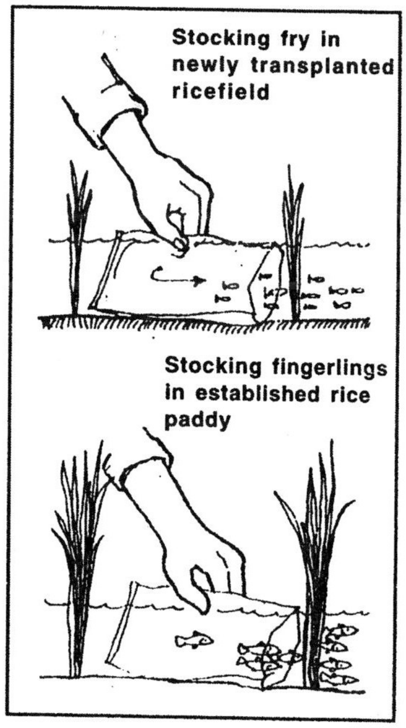

- Pesticide application

  Similar to fertilizer application, damage to fish health can be incurred with the application of pesticides to the rice crop. However, using simple techniques such as slow field drainage, allowing the fish to return

to the trench/pond, cautious application of the pesticide, an allowance for a brief waiting period and re-irrigation of the field after application can help ensure minimal losses due to pesticide poisoning.

In fields with trench/pond designs, the water should be drained into the trench or pond, thus driving the fish into the refuge area before the application of pesticides.

In traditional paddy field design, fish can be driven to one-half of the field and pesticide application can be done in the other half of the field. The same procedure can be repeated to the other half of the field on the following day.

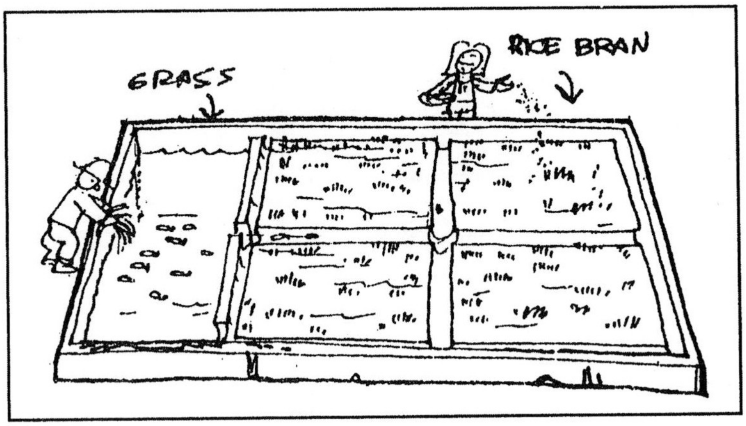

Issues for further consideration

With increased intensification in rice farming in China, the importance of azolla is likely to have declined. Which variability exists for different systems, i.e. rainfed, irrigated, traditional and improved, and what is the potential for further expansion? The market patterns of produced fish should be known before embarking on such an activity. Information on applications of nursery or growout systems.by different types of farmers and the relevance of the system within their overall farm would provide important pointers for further applications.

The example provided on pesticide application in a traditional paddy field design only applies to chemicals which are not toxic to fish.

Rice-prawn culture in the Mekong Delta of Viet Nam

Le Thanh Duong

Rice-prawn farming, a traditional practice in freshwater areas of the Mekong Delta, has become more popular for the past few years. This paper is based on two rice-prawn farming systems tried in one of the key sites (Phung Hiep district, Han Giang) of the Mekong Delta Farming Systems Research and Development Centre as part of an OXFAM-funded project. In 1990, the following farming systems were tried in the site:

- System 1 -- prawn integrated in dry season modern variety rice-wet season, modern variety rice (DS MVR-WS, MVR).
- System 2 – prawn integrated in wet season modern variety rice-wet season transplanted rice (WS MVR-WS TPR).

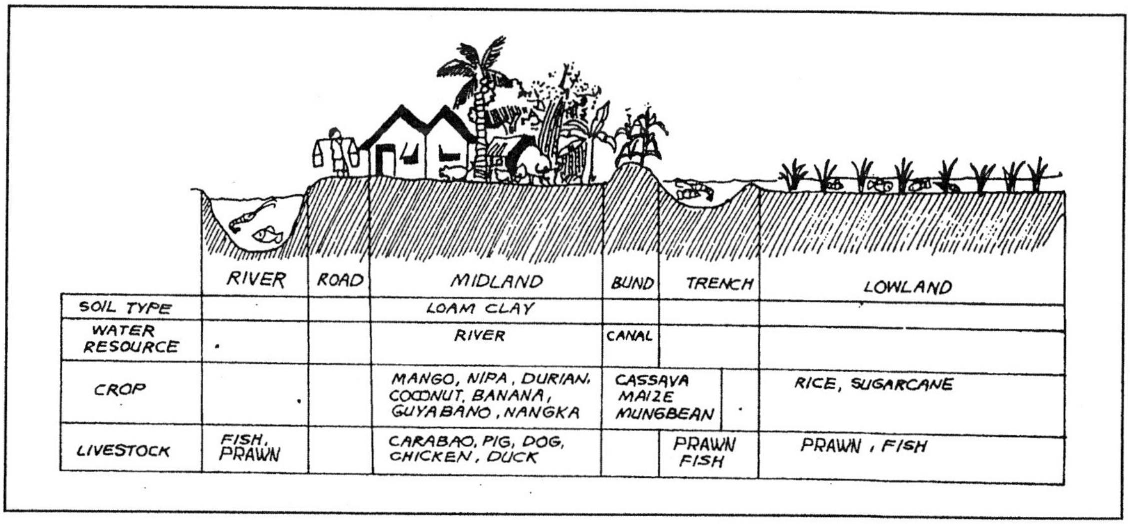

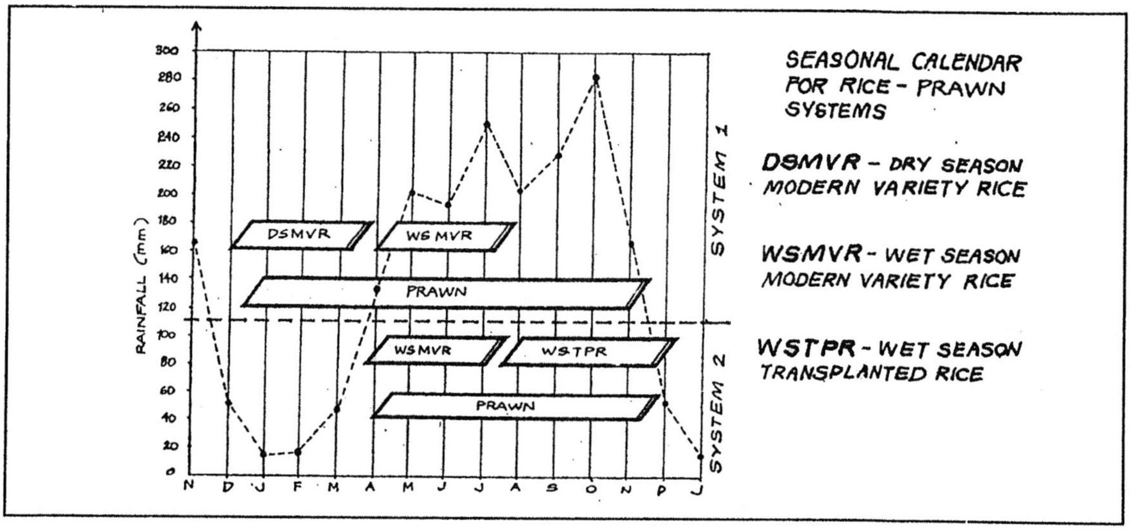

Culture practices

The following are the eight main steps in rice-prawn culture:

1. Field selection
   - The field should be close to a water source and should not have salinity or acidity problems. Water depth can range from 20 to 30 cm.
   - Flat fields are often preferred with an average of 1-2 ha. Peripheral dikes should be high enough to prevent flooding in the wet season and should be compacted so that water does not leak and percolate in the dry months.
2. Land preparation
   - Dig trenches inside the dikes on all four sides of the field, 3-4 m wide and 1-1.2 m deep. Make supplementary trenches outside the dikes to store prawn juveniles or adult prawn as needed.
   - Make at least three sluice gates with 0.3-0.4 m diameter. Two are installed at 0.2 m above field level to retain water and one at the lowest level of the trench for draining during harvest time. Put a net or woven strips of bamboo on the sluice gates to prevent fish and prawn from escaping.
   - Prepare the field thoroughly before stocking. Deposit materials and mud should be removed.
   - Apply powdery lime (100 kg/1 000 m^2) or *Derris elliptica* roots (1-1.5 kg soaked in 10-15 litres water/1 000 m^2) to help get rid of wild fish and other carnivorous animals (e.g. crabs, snakes, frogs, etc.)
   - Sun-dry trench bottom for 3 days so that it becomes solid. This prevents mud from penetrating into the prawn's filament chamber and feeds from sinking into the mud.
   - Cover 8-10 percent of the water surface in the trench with plant branches to discourage poaching.
3. Selection of prawn juveniles and rice seed varieties
   - Gather healthy juveniles of freshwater giant prawn (*Macrobrachium rosenbergii*) from the river. They should be confined in supplementary trenches or stocked immediately if the trenches are already prepared.
   - Short-duration modern varieties can be used, such as IR66, MTL85, MTL86, MTL88 (100-110 days) or medium maturity modern rice or high-yielding local varieties such as *Mot bui lun, Lua thom, Trang tep, Tai nguyen* (which flower on 20-30 November). Use seed stock with more than 80 percent germination rate.
4. Rice crop establishment and prawn stocking
   - Stock with juveniles (size of 100-250/kg) at the rate of 7-8 kg/ha.
   - Put the juveniles in a big bamboo basket and slowly submerge this into the water. Stocking should be done after land preparation or rice broadcasting or transplanting.
   - If two modern rice crops are grown: Broadcasting or transplanting time for dry season crop is in November-December. Prawn fry are stocked in December at 10 days after broadcasting or 5 days after transplanting. Prawn will be stored in the trenches during harvesting or dry season crop or during land preparation for wet season crop.

     After broadcasting/transplanting wet season rice crop in March-April, allow prawn into the ricefield at 10 days after broadcasting or 5 days after transplanting.
   - In wet-season modern rice-transplanted local ricefields: Prawn juveniles are also stocked in December in supplementary trenches for

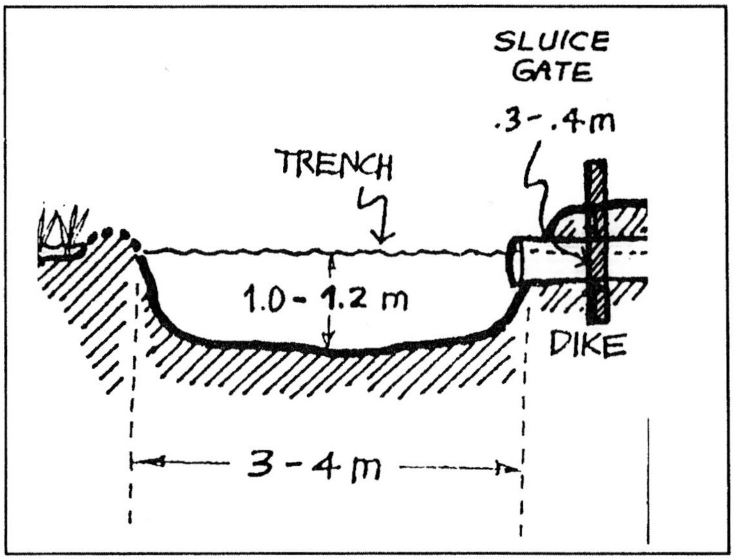

Material flow in rice-prawn integrated farming systems of Mekong Delta, Viet Nam

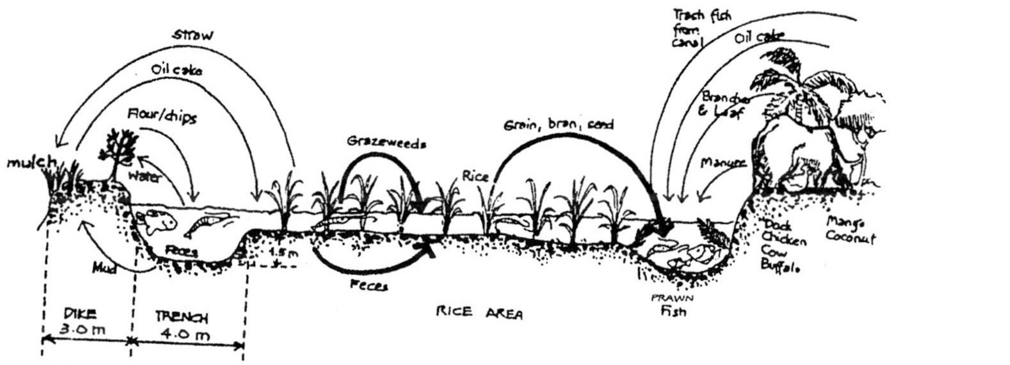

storage and these can be released into the field in March-April (10 days after broadcasting or 5 days after transplanting of wet-season crop) and in July-August (10 days after broadcasting or 5 days after transplanting of transplanted local rice). They can be placed in supplementary trenches during harvesting of wet-season crop (June-July) or land preparation for transplanted local rice crop (July-August).

5. Feeding, weeding and fertilizing
   - Prawn feeds
     - Cassava, sweet potato, broken rice, milled rice (soaked or cooked), rice bran
     - Crabs, snails, trash fish
     - Copra
     - Combined feeds: 50 percent rice bran, 10-20 percent cooked broken rice, 20-30 percent trash fish, 10 percent oil cake
   - Feeding
     - Feeding ratio: 3-5 percent of prawn weight
     - Feeding times: at least twice a day (1/3 at 5-6 a.m. and the remaining 2/3 at 5-6 p.m.)

     Put the feeds in feeding trays and place them anywhere along the bottom of the trench.
   - Weeding. Hand-weeding is recommended at 15 and 35 days after broadcasting or 15 and 30 days after transplanting. Herbicides should only be applied if necessary.
   - Fertilizing. Both organic and inorganic fertilizers are used. Organic fertilizers and phosphorus can be basally incorporated in the soil. Nitrogen should be split (divided between two applications). Complete fertilizers are applied at 10 days after broadcasting or split for transplanted rice. Potassium can be basally applied and topdressed.

     Fertilizer formula for modern rice per hectare:
     - 200 kg monosuperphosphate + 200 kg urea + 50 kg potash
     - or 100 kg 18-46-0 days after planting + 100 kg urea + 50 kg potash

     And for transplanted local rice per hectare:
     - 200 kg monosuperphosphate + 100 kg urea + 50 kg potash
     - or 100 kg days after planting + 50 kg urea + 50 kg potash

6. Management of water, sluice gates and dikes
   - Change water every day with the tidal regime. Water level in the field is kept at more than 20 cm and at 100 cm in the trenches.
   - Check water quality every day. If it is acidic or hot, change the water or add more water into the field.
   - The water surface of trenches should be exposed to air; only 15-20 percent of surface may be covered by floating weeds or vegetables like water spinach, etc.
   - Check surrounding dikes regularly for leaks. Put fish nets or bamboo fences on sluice gates to prevent prawn from escaping.

7. Other management practices
   - An advanced management practice is to record growth rate of prawn monthly by weighing them. Average growth rate is about 5-6 g/prawn/month. If growth rate falls below 3 g, growing conditions should be improved or more feeds should be added.
   - Prawn molt about twice a month. After each molting, prawn weight will increase from 3 to 5 g. Molting often occurs in the early morning or at night time during low tide.
   - If water is deficient in oxygen, prawn often appear at the water surface in the early morning. When deficiency is more serious, most of the prawn may die. Maintain oxygen levels by keeping recommended water depth and feed regularly to prevent water from getting polluted.
   - Some minor diseases can hamper the prawn's growth rate. Improving growing conditions (e.g. using powdery lime before prawn stocking, keeping water clean during the growing season), using good fry and controlling parasites normally suffice to prevent diseases.
   - Control carnivorous fishes and other animals.

Average prawn and rice yields in the 1990 crop season

Rice crop	Yield (t/ha)	% of farmers getting >5 t/ha	4-5 t/ha	<4 t/ha		Prawn in systems (kg/ha) Average	Maximum	Minimum
DSMVR	4.7	31	51	18	2MV	97.5	285	24
WSMVR	5.4	25	40	35	MV-TPR	98.0	364	13
WSTPR	4.3	27	30	43				

Legend:
DSMVR : dry season modern variety rice
WSMVR : wet season modern variety rice
WSTPR : wet season transplanted rice
2MV : dry season modern rice followed by wet season modern rice
MVTPR : wet season modern rice followed by wet season transplanted rice

Partial budget (in thousand VND/ha) for two dominant rice-prawn systems

	MV – MV – Prawn Quantity	Cost return	MV – TPR – Prawn Quantity	Cost return
Gross return				
Rice	3 875 kg	3 488.0	5 880 kg	6 467.9
Prawn	98 kg	2 100.1	98 kg	2 135.9
Costs				
Rice : hired labour	124 person-days	496.4	143 person-days	574.0
materials		679.4		503.1
others		84.0		84.0
		1 259.8		1 161.1
Prawn: hired labour	34 person-days	134.4	51 person-days	203.9
materials		638.9		790.6
Subtotal		773.3		994.5
Net return of rice		2 228.2		5 306.8
Net return of prawn		1 326.8		1 141.4
Total net return		3 555.0		6 448.2

1992: US$1 = 7 000 VND
This table shows partial budget of rice monoculture and of prawn integrated in the two dominant rice cropping systems. Net return from prawn in the two systems seems to be the same while net return from WSMV/TPR is higher than that of 2MV system. Prawn contributed significantly to net income of rice farmers in the site.

- Use rice varieties resistant to major insects and diseases to minimize use of chemicals. If chemical application cannot be avoided, partially drain the field so that the prawn can take refuge in the trenches. Water should be changed completely 3-4 days after pesticide application.

8. Harvesting of rice and prawn
   - Prawn are harvested in November/December before harvesting of transplanted local rice crop and before land preparation for dry-season rice crop. They can also be partially harvested after 4-5 months of growing. Only the larger ones will be taken and the rest are restocked with an additional amount of the same-size prawn.
   - Rice can be harvested when 80 percent of the crop is mature. Delay in harvesting can result in serious grain shattering. After threshing, rice should be sun-dried and stored.

Issues for further consideration

The system described is characteristic for the Mekong Delta and other areas where wild or hatchery-reared *Macrobrachium* juveniles are available and rice production is of moderate intensity. The existing rice varieties and culture systems (e.g. wet season, modern varieties, transplanting, etc.) and options for modifying these will influence the design of an adoption.

With a harvest of 13-24 kg/ha, questions arise as to the continued viability of these systems. Different options exist to improve these systems. Associated risks to the farmers will need to be considered.

In assessing economic viability in given local situations, additional rice fertilization due to more frequent water exchanges, feed costs for prawns, costs of other off-farm materials such as trash fish, have to be included.

Rice-prawn and rice-shrimp culture in coastal areas of Viet Nam

Le Thanh Hung

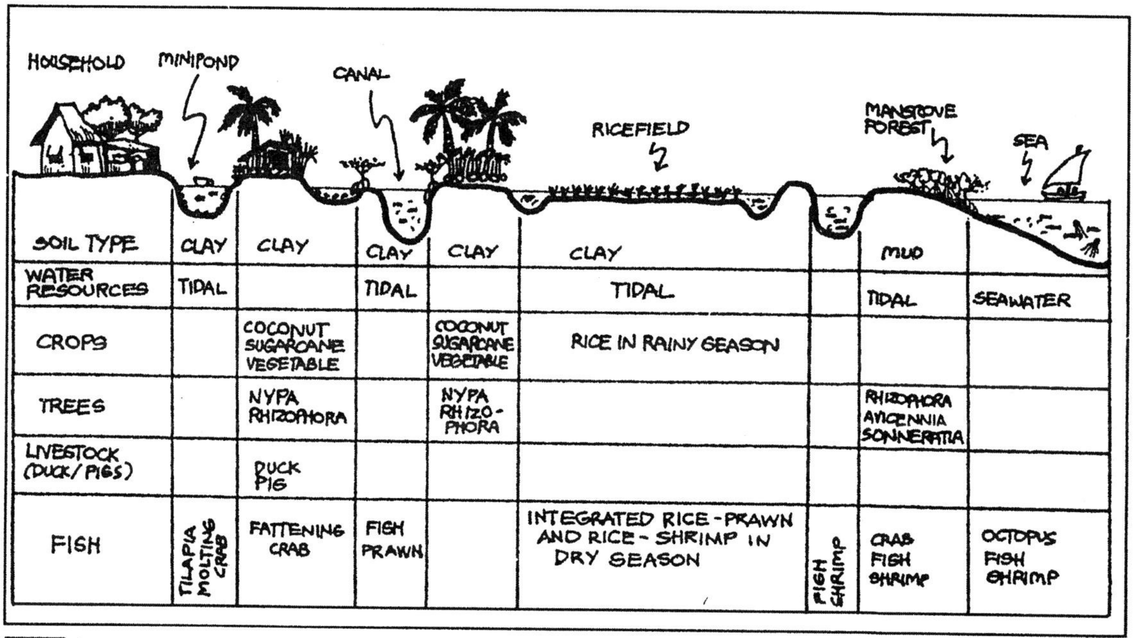

Tidal flats in coastal areas are flooded periodically during high tides. During the dry season, salinity is usually higher than 5 ppt (per thousand); thus, most paddy fields are in fallow. In the rainy months, salinity declines so rice cultivation is possible. Farmers in these coastal areas of southern Viet Nam have living standards lower than their counterparts in freshwater regions. Integrating freshwater prawn culture with rice during the rainy season, as well as marine shrimp monoculture in the dry season, is one way of increasing their incomes. Both systems are described here.

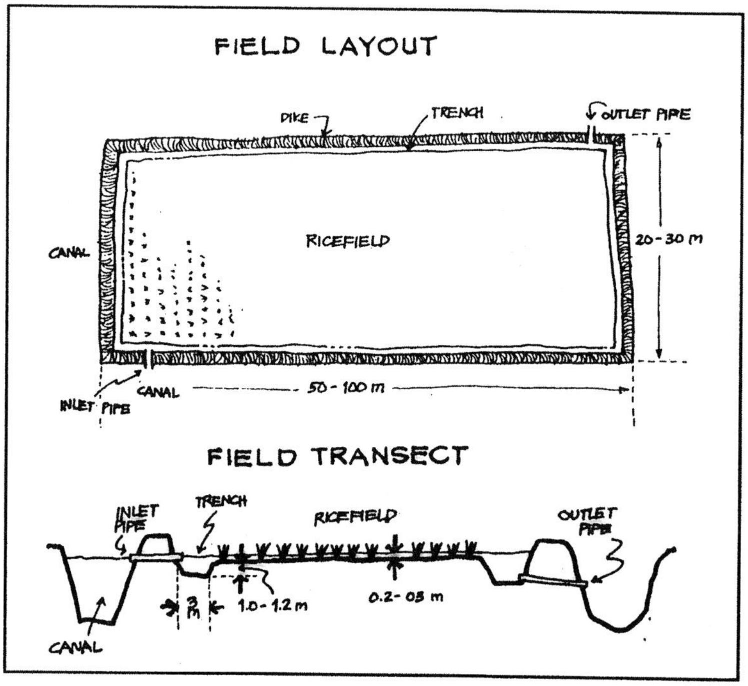

Site selection

The field should be close to a river or channel.

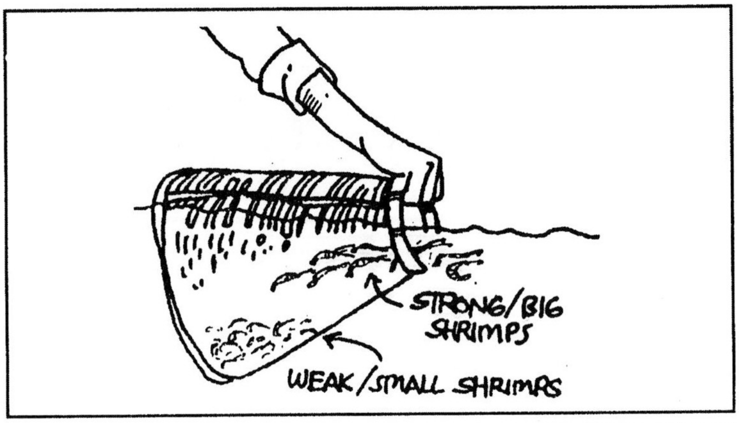

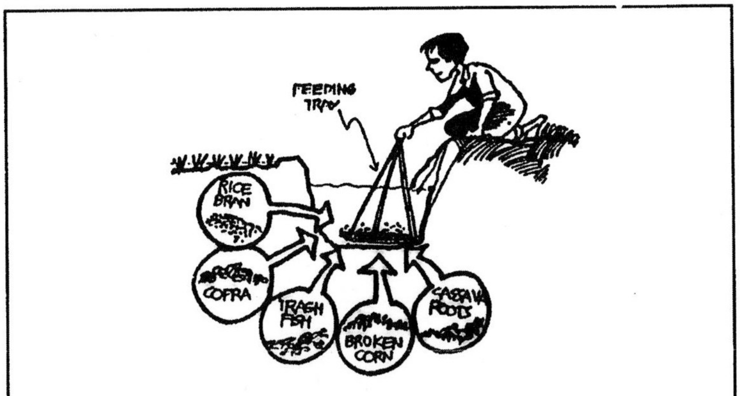

	MONTH	1	2	3	4	5	6
5%	FEEDING FOR 1000m² (kg)	0.5	0.6	0.8	1.0	1.2	1.5
2-3%	1000m(kg)	0.3	0.4	0.5	0.6	0.8	1.0

- Choose a low and flat place so it is easy to get water during high tide.
- Avoid high acid sulphate soils.

Dike and trench construction

- Surface area of field: 1 000-3 000 m²
- Trench is 2-3 m wide and 0.8-1.0 m deep with trench-to-ricefield rate of 10-20 percent.
- Peripheral dikes should be at least 20 cm higher than the annual flooded level.
- Install 2-3 inlet and outlet pipes (at least 20 cm diameter) made of coconut trunk or wood. The inlet pipe should be installed so as to let water into the paddy field at high tide; the outlet pipe should allow water to drain from the trench when opened.
- Inlet and outlet pipes should be screened to prevent the intrusion of predators.
- Cover the trench surface with tree branches or plant water hyacinths, etc., along the trench to discourage poaching.

Stocking

- Stock juveniles of giant freshwater prawn *(Macrobrachium rosenbergii)* at a density of 1.2/m² (at least 4-5 g each).
- Stock 10-15 days after transplanting.
- Criteria for juveniles: vigorous, strong and uniform in size

Note: If stocking density is higher than 1/m², supplementary feeding should be done and trench-to-ricefield rate should be higher than 10 percent. If water exchange is poor, do not stock higher than 1/m².

Feeding

- Prawn can subsist on natural food in the paddy field, especially if it is loaded with manure.
- The following supplementary feed can be given: rice bran, rice grain, copra, oil cake, cassava root, broken maize, fiddler crab (*Uca* spp.), shrimp or prawn head wastes and trash fish.
- Feeds can be given daily at 5 percent of the prawn's body weight (if no manure loading) or 2-3 percent (with manure loading). Mix ingredients thoroughly, form them into balls and put them in feeding trays. The use of feeding trays controls consumption of feeds and prevents wastage.
- Feed twice a day: one-third of the quantity in the morning and the rest in the afternoon.
- Check feed consumption daily to adjust the feeding regime as necessary. Below is a recommended formula for prawn in rice paddies.

50 percent – rice bran, broken rice or rice grain
20-30 percent – cassava root or broken maize
20-30 percent – trash fish, shrimp or prawn head wastes or oil cake

Predator prevention

Predators include sea bass, tilapia, snakehead and other wild fish that compete with the prawn for feeds. Predation can result in very low prawn yields.

Before stocking prawn, use any of the following measures:

- Drain ricefields and apply lime at the rate of 10 kg/100 m² (15-20 kg for acid sulphate soils).
- Apply Derris root (*Derris elliptica*), 1-1.5 kg soaked in 10 litres water/1 000 m².
- Release ducks into the

ricefields for several days.

Within the culture time, put gill nets in the trenches to catch the predators going to the ricefields.

Care and maintenance

- Water exchange is essential to supply oxygen to the prawn and to remove detrimental substances in the water. This should be done at least twice a month. The more frequent the water is changed, the more suitable it is for the prawn's growth and development.
- Water exchange also improves the pH value in the fields especially in sulphate acid soils.
- Dikes should be repaired yearly.
- Cover crab holes along the dikes to prevent leakage.
- Check daily the screen mesh on the outlet and inlet pipes.

Harvesting

- Harvest prawn 5-6 months after rice harvest.
- Open the outlet pipe at low tide and drain the field and trench.
- Hand-collect prawn in the ricefield and use a net to harvest in the trench.
- Harvest only the big (more than 15 g) prawn. The small ones are reserved for the next culture season.

Note: Transfer small prawn immediately to a hapa (net cage) to keep them alive for the next culture. Bring harvested prawn as soon as possible to the dealer or keep them in ice so that they stay fresh.

Land preparation and transplanting for rice

- Local varieties are recommended. Transplanting should be done when the salinity is lower than 5 ppt.
- Plough and harrow thoroughly before transplanting.

- Transplant 3-40 days after seeding.

Fertilizing

- Apply 50 kg diammonium phosphate and 5 t manure/ha before ploughing.
- Use 50 kg urea/ha for top-dressing.

Pest control

- No pesticide or herbicide is applied in integrated prawn-rice culture.
- Use brown planthopper-resistant varieties of rice.
- Release one-month old ducks into ricefield to feed on insects, especially hoppers.

Note: In case the above measures cannot control pests, pesticide application can be an alternative. Before applying the pesticide, drain water in the field to let prawn take refuge in the trench for 3-5 days.

Monoculture of prawn or shrimp in fields during dry season

- Nipa and coconut trees are indicators of salinity lower than 10 ppt. *Rhizophora* (a mangrove species) is an indicator of salinity higher than 10 ppt.
- During the dry season, when salinity level is high and therefore not suitable for rice growing, the fields can be used for shrimp monoculture.
- Freshwater prawn *(M. rosenbergii)* can be grown if the salinity is not higher than 10 ppt. The procedure is similar to that applied in the rainy season. When the salinity is higher than 10 ppt, freshwater prawn become stunted.

- Tiger shrimp *(Penaeus monodon)* and banana shrimp *(P. merguiensis)* can be cultured in ricefields when the salinity is higher than 10 ppt in the dry season.

Stocking density - 1/m²
Stocked juveniles - 2 g/head
Feeding rate - 2-3 percent of body weight
Feed formula - 50 percent rice bran (broken rice), 50 percent trash fish (fiddler crab, oil cake)
Culture time - 5-6 months

Other procedures are similar to freshwater prawn culture.

Estimated cost and return (in VN$) of rice-prawn culture in 1 ha coastal areas in South Viet Nam

Costs	
Rice seed (24 kg x VN$2 500)	60 000
Prawn seed (4 000 seeds x VN$60)	240 000
Fertilizer (20 kg urea x VN$ 2 500)	50 000
Feed (100 kg x VN$1 000)	100 000
Manure (1 t x VN$20 000)	20 000
Miscellaneous	20 000
Labour (20 persons/day x VN$7 000)	140 000
Total cost	630 000
Income	
Rice (0.5 t x 1 200 000)	600 000
Prawn (30 kg x 35 000)	1 050 000
Wild fish (4 kg x 5 000)	20 000
Total income	1 670 000
Balance	1 040 000

Issues for further consideration

The situation described here is that before 1992 when the major shrimp disease outbreak in Asia occurred. Since then, disease in shrimp farming has become a major risk that has totally altered the economics of the system and has to be considered according to local prevalence. It is a major constraint to farmers, even for very extensive operations. Existing and ongoing epidemiological studies by Stirling University is applying epidemiological approaches to understand risk factors for farmers in these systems.

Resource constraints to these systems should be given attention. All the feeds given represent an investment in time or cash and this should be considered. Collection of snails and fish is a major activity that is not typically shared equally by all household members.

There are large areas of semi-saline zones around Asia (e.g. Bangladesh and India) and elsewhere, and this case study should be of relevance there. These tend to be marginal areas where poorer people live.

In other contexts, after shrimp production became established, outside interests have become influential and extract much of the value from shrimp culture. In Viet Nam, government policy prioritizes increased rice production areas. Large areas are being protected from saline intrusion, and converted to triple rice crop production, from previous shrimp culture, although in the transition, farmers are trying to grow shrimp as long as possible (e.g. grown at salinities less than 4 g/litres) due to much higher returns.

Rice-fish system in Guimba, Nueva Ecija, Philippines

Catalino dela Cruz, Ruben C. Sevilleja and Jose Torres

Guimba, Nueva Ecija, Philippines, has rainfed and irrigated rice-based agriculture. In rainfed areas, rice is grown during the wet season and remains fallow during the rest of the year. In irrigated conditions, rice grown during the wet season is followed by another crop of rice during the dry season. Rice-fish culture is practiced by some farmers. In areas with extremely light soils, farmers plant vegetables (e.g. squash, cucumber, mungbean, stringbeans, onions, bitter gourd, etc.) and watermelon after wet season rice.

The rice-fish system practiced by farmers in Triala Village, Guimba, is concurrent rice-fish with pond refuge and is irrigation-based. This system is for growout of Nile tilapia. The operation is done as follows.

Rice-fish field design and construction

1. Site selection
   - Abundant and dependable water supply. Irrigation water, ground water, spring and other water sources are used when they are not contaminated by pesticides.
   - Clay soil is best. Clay holds water, prevents seepage and leaching of fertilizers.
   - Choose site with good drainage and that is free from flooding.
2. Design and size of field
   - Independent filling and draining of each rice-fish compartment is considered.

Supplemental feeding of fish in refuge of rice-fish field

   - Ease of fish movement into the ricefields during grazing and draining is also considered. The fish should be able to get quickly into the canals or refuge when water level is very. low.
   - Size of rice-fish plot considers the natural partitions of the field. Small plots are easy to manage, and fish survival is usually high.
   - Dikes are made strong and big enough to withstand the pressure of water level in the flooded ricefield.
3. Fish refuge

Pond refuge, which holds more water and is less risky, is preferred over trench refuge. Refuge size is usually 10 percent of the ricefield area. Bigger refuge or a pond adjacent to the ricefield may also be connected to it through a canal.

To construct the refuge, the pond is excavated at one end, or two ends if the field is large, inside the ricefield or adjacent/alongside but connected to the field so that the fish can have access to the area planted to rice.

Examples of fish refuge layout for small and large plots

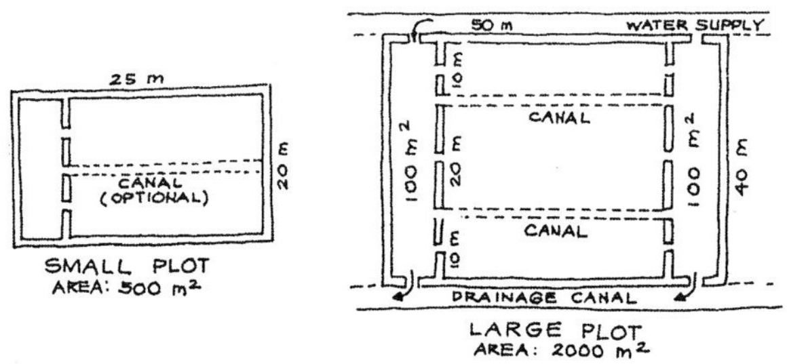

4. Inlet and outlet gates and screens
These are made of bamboo and other low-cost materials. Screens prevent the escape of stocked fish or entry of unwanted fish into the field.

Rice agronomy

1. Rice varieties - high-yielding varieties; maturity period of 120-130 days; resistant to insects and diseases.
2. Seedbed preparation and seeding rate

Size	: 400-500 m²
Rate	: 100-150 kg/ha
Fertilization	: broadcast urea at 25 kg/ha 10-15 days after sowing

3. Land preparation - After ploughing once and harrowing thrice, the field is leveled evenly so that every part of it will be uniformly irrigated.
4. Rice transplanting method
   Age of seedlings : 25-30 days
   Planting distance :
   20-25 cm between rows
   15-20 cm between hills
   Straight-row planting (optional), if mechanical weeding is done.
5. Weed control - Fish stocked in ricefields control certain weeds. Weeds are also controlled through:
   - thorough land preparation;
   - flooding the field at an effective water depth for 1-2 weeks immediately after transplanting; and
   - manual weeding.
6. Water management - Water depth in the field when rice is newly transplanted is 3-5 cm. This is then gradually increased up to 20 cm to provide better living space for both rice and fish as they grow bigger. One week before the rice harvest, water is slowly drained so that fishes have enough time to move into the refuge.
7. Fertilizer application - The amount of fertilizer applied follows the recommended rate in the area. In Guimba, the rate applied during wet season is 200 kg/ha of ammonium phosphate and 50 kg/ha of urea for the first or basal application. The basal application is done immediately after the final leveling, which is followed by transplanting. The rate for the second application or top-dressing is 50 kg/ha. This is applied 30 days after transplanting. The amount for top-dressing may be split into two equal applications. In this case, a third application is applied 75 days after transplanting.

   During dry season, the same amount for basal application is followed. For top-dressing, the rate is 100 kg/ha. As an example, the amount of fertilizer for a 400 m² of rice-fish during wet season is 8 kg of ammonium phosphate and 2 kg of urea for basal application. For top-dressing, 2 kg of urea are needed.
8. Insect control - The use of insecticide is not recommended. Farmers, however, apply insecticides known to be less toxic to fish.

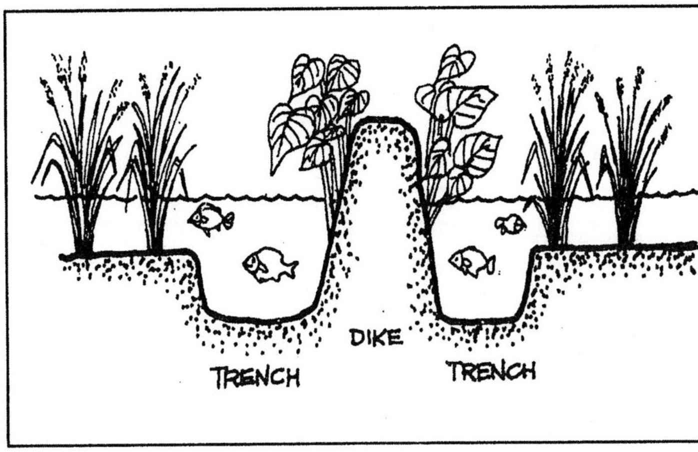

Raising *Colocasia*

Colocasia sp., an aquatic plant also known as taro, is an excellent food material. It can be grown as an added commodity in a rice-fish farm. Practically, all parts of the plant can be eaten (tubers, stalks and leaves). It can also be utilized as food for fish and animals, especially pigs. The culture requirement is simple, and there are no expensive inputs.

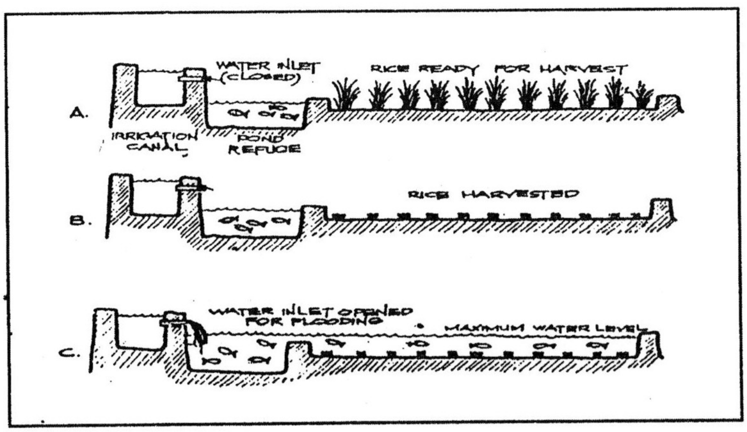

taro

Here are steps in raising *Colocasia:*

1. Obtain young tubers.
2. Cut old leaves but retain the young leaves and shoot.
3. Cut the tuber into half.
4. Plant the tuber at 50-70 cm intervals along the side of the dike, about 10 cm below the water surface.
5. Start harvesting after 4-5 months.

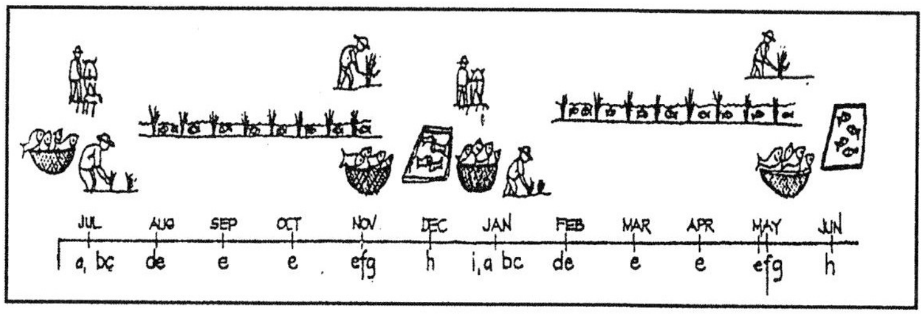

Activity calendar for rice-fish culture

Code	Day	Activity
a	0	Prepare and fertilize seedbed.
a	1	Soak seeds (IR-36, 42, 52, 64 and 74 as examples).
a	3	Broadcast germinated seeds in seedbed.
b	7	Prepare ricefield. Start of fish culture (Stage 1). Stock fish (Nile tilapia, 5-10 g in size) at 5 000-7 500/ha. Insure water supply in fields.
c	24	Pull seedlings. Apply basal fertilizer. Use the kinds and rates of fertilizer recommended for the locality based on soil analysis.
d	25	Transplant seedlings (wet bed method).
e	51	Apply fertilizer (top-dressing) for the second time; may split the amount into two applications, hence, there may be a third application.
e	75	Apply fertilizer (top-dressing) for the third time.
f	100	Drain water and harvest large/marketable fish.
g	12C-125	Harvest rice.
g	125	Start of fish culture (Stage 2). Prolong fish culture period after rice harvest for small-size fish (30-40 g) stocked at 3 000-5 000/ha.
h	185	Harvest second batch of fish.

Fish culture

The species cultured are Nile tilapia *(Oreochromis niloticus)* and common carp *(Cyprinus carpio).* Large fingerlings, 15-25 g, are recommended as they reach harvestable size within one rice crop. If only small fingerlings (5-10 g) are available, fish culture is done in two stages:

Stage l: Raising 5-10 g fingerlings during one rice cropping (harvest size: mostly 50 g)

Stage 2: Extending fish culture period after rice harvest for up to 2 months (harvest size: 50 g)

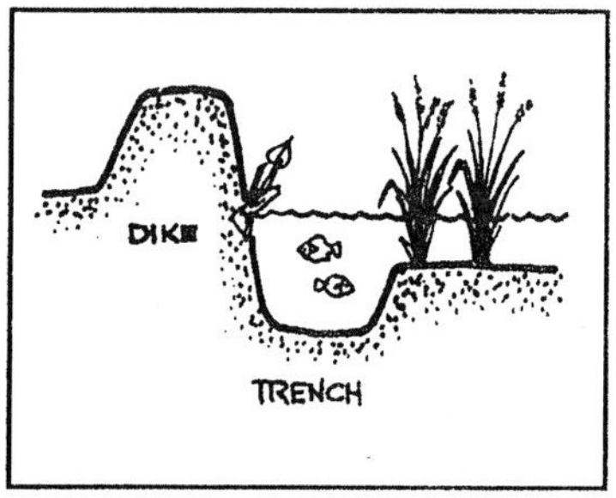

Stocking density

- Stocking can be done before or during land preparation in the pond refuge; or 7-10 days after transplanting (DAT), if fish are released directly to the fields. If stocked in the pond refuge, animal manure should be applied into the refuge 4-5 days before fish stocking. About 15 kg may be applied in a 100 m² pond refuge.

  The stocking rate for Stage 1, using either monoculture of Nile tilapla or polyculture of Nile tilapia and common carp is 5 000-7 500 fish/ha. For polyculture, the stc:king ratio of Nilc tilapia to common carp is 1:1 or 2:2, depending on which species is more important to farmers.
- Ten days after transplanting, fish stocked in the pond refuge may be released to the field by making openings in the dividing dike. Fish will graze on natural food available in the ricefield.

Supplemental feeding

- This feeding is recommended at the middle culture period of rice (45-50 DAT), during which production of natural food in the fieldwater declines due to shading of rice leaves.
- Feeds: rice bran, kitchen refuse, ipil-ipil meal, etc. Animal manure may also be applied in the pond refuge.
- Feeding rate: 3-5 percent of fish biomass

Harvesting

- Harvest fish by draining the water very slowly 1 week before rice harvest to avoid trapping the fish in the middle of the field.
- Select large fish for consumption or disposal and confine the small fish (50 g) for stage 2 culture.
- After harvesting rice, the field is immediately reflooded to about 30 cm deep, and the small fish in the refuge are released to allow them to grow for another 60 days before the dry season crop.

Benefits and limitations of the system

1. Fish can contribute to increased rice yield by 10-15 percent by:
   - Controlling certain weeds and insects such as stemborer and brown planthopper;
   - Producing fish wastes, including uneaten feeds which add fertility to the soil;
   - Increasing availability of nutrient for increased floodwater productivity and uptake by rice; and
   - Reducing loss of ammonia through volatilization by pre-

Annual budget (in US$) for a 1 ha rice-fish farm with pond refuge

Item		Amount		
I. Returns		Rice-fish		Rice-fish + taro
Rice (2 crops)		1 457		1 457
Fish (2 crops)		386		386
Taro (2 crops)		-		581
Total returns		1 843		2 424
II. Costs				
Labour		402		515
Materials		375		428
Seeds	50		50	
Fingerlings	41		41	
Taro tubers	-		53	
Fertilizers	140		140	
Pesticides	37		37	
Fuel and oil	86		86	
Feeds	16		16	
Others (screens, bundling materials, etc.)	5		5	
Total costs		777		943
III. Net returns		1 066		1 481

Partial budget (in US$) for integrating taro with rice-fish farming (1 000 m² area)

Added costs		Added returns	
Labour (Planting and harvesting)	5.65	Taro	31.70
Materials (Tubers for planting)	5.30		
Reduced returns		Reduced costs	
Total	10.95		31.70

Note: Increase in net income: 31.70 - 10.95 = 20.75

venting floodwater pH to rise over 8.5. During fertilizer application, increased plankton production tends to raise the value of pH beyond 8.5, the value at which ionized ammonia converts into an un-ionized form that is easily lost.

2. The increased size of dikes in the system offers opportunity to plant other crops, such as taro (*Colocasia* sp.), stringbeans, cowpea, wingbeans, eggplant and others.
3. The wide-scale adoption of rice-fish is still constrained by continued application of pesticide in rice-based farming. The use of pesticide is not recommended in rice-fish farming. There are ways of controlling rice pests that do not need pesticide, such as:
   - Quick submergence (for 3 hours) of rice plants in water. This makes the insects vulnerable to fish predation. Limitation: suitable while rice plants are shorter than the dike.
   - Two persons can drag a stretched rope (50-100 m) across the ricefields to knock off the insects into the floodwater, after which they can be eaten by the fish. Limitation: suitable before rice plants reach panicle initiation stage.

However, should a farmer insist on using pesticide, here are ways to apply it:

- Choose and apply pesticides that have low toxicity to fish properly.
- Minimize the amount of pesticide getting mixed with water.
- Apply at suitable time.

• Preventing fish poisoning:
- Drive the fish into the refuge by draining the field before spraying. Keep the fish in the sump until toxicity in the sprayed field is gone.
- Increase water depth (+10 cm) to dilute the concentration of pesticide in the water.
- Flush water through the ricefield. Open the inlet and outlet of the field and allow irrigation water to flow freely, during spraying. Begin spraying from the outlet end of the field. When one-half of the field is already sprayed, stop for a while and allow the pesticides to flow out of the field. Then, continue spraying towards the inlet end of the field until it is finished.

To do the last two items above in applying pesticides, here are some examples: apply powder pesticides in the morning when dewdrops are still on the leaves; and apply liquid pesticides in the afternoon when leaves are dry.

There are a number of less toxic pesticides in the market Proper application of a toxic insecticide like Furadan ® or Curaterr ® can be made safe to fish if applied through soil incorporation during the final harrowing. Furadan ® is a systemic insecticide, the efficiency of which in controlling insect pests lasts about 50-55 days. Incidence of pests after this period can be controlled by spraying liquid pesticides. At this time, the rice plants have reached their full vegetative stage and the thick leaves will intercept most of the liquid sprays, thus, drastically reducing the concentration of pesticide reaching into the water.

Issues for further consideration

Rice-fish farming was adopted by a number of farmers in the area, but not many of these have sustained their operations. The taro (i.e. *Colocasia*) component may be regarded as a cash or subsistence crop. Its value may vary differently by season than rice. With higher fish stocking rates as described here, supplementary feed will be necessary.

Today, integrated pest management (IPM) is the declared national pest management strategy in the Philippines and many other rice-producing countries, and the International Rice Research Institute has published results that natural control of rice pests without pesticide use generally is the most profitable option for rice farmers. The concept of IPM certainly precludes the use of systemic insecticides for preventive treatment.

The case of rice-fish farmer Mang Isko from Dasmariñas, Cavite, Philippines

Frank V. Fermin, Mary Ann P. Bimbao
and Jens Peter Tang Dalsgaard

Household profile

Mang* Isko is a 66-year old farmer. Together with his wife, who is 60, they have eight children most of whom are grown-ups and living away from home. The only son is married and living with his wife and children near the farm of Mang Isko. This son helps Mang Isko in the day-to-day management of the farm. Two daughters are attending high school and still live at home. Two older daughters, who are working in Japan, send P4 000/ month to support the education of their younger sisters.

* "Mang" is a respectful address for an elder in the Philippines

Mang Isko's farm transect (lowland farm)

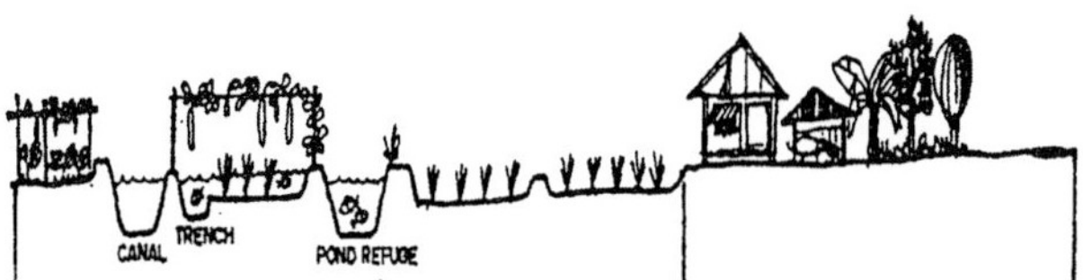

SOIL	SANDY CLAY LOAM	CLAY LOAM	SANDY CLAY LOAM (WITH FERTILE TOPSOIL FROM LITTER)
WATER	RAIN AND IRRIGATION		
CROPS AND VEGETABLES	BITTERGOURD, STRINGBEANS, CHILI, SWEET PEPPER, RICE-FISH, GOURD	RICE	TOMATO
TREES			MANGO, BANANA, GUAVA, CALAMANSI, PAPAYA, LEUCAENA
FORAGE			PARAGRASS
ANIMALS/ FISH	TILAPIA AND CARP		PIG, CHICKEN

Mang Isko's on-farm material flows

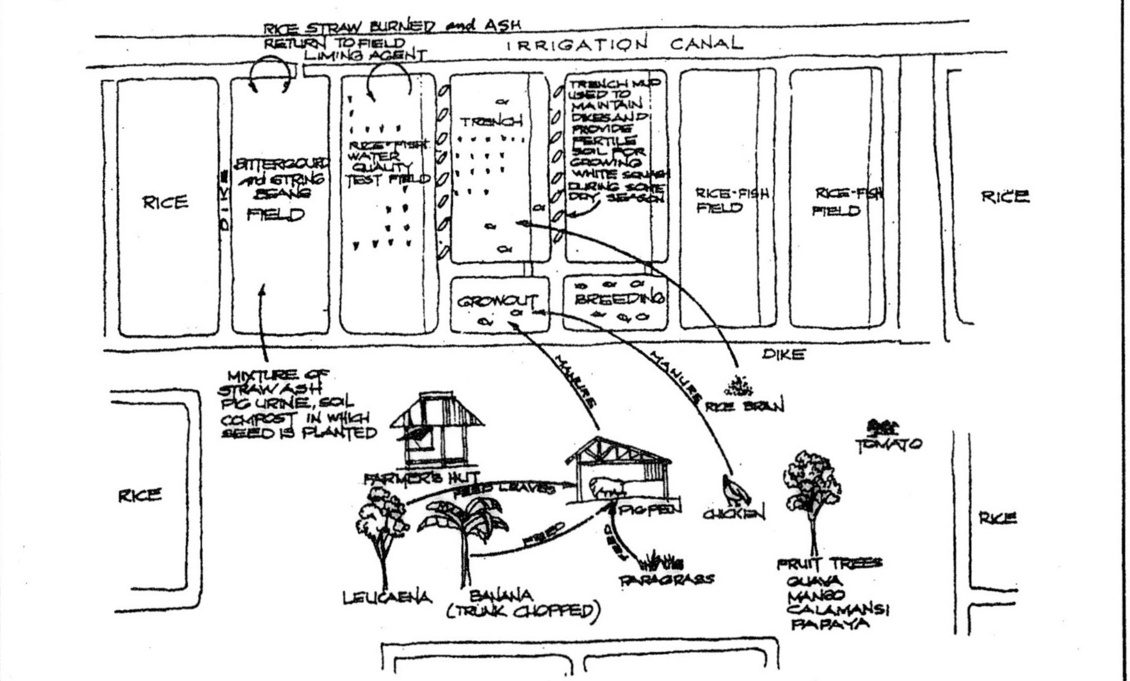

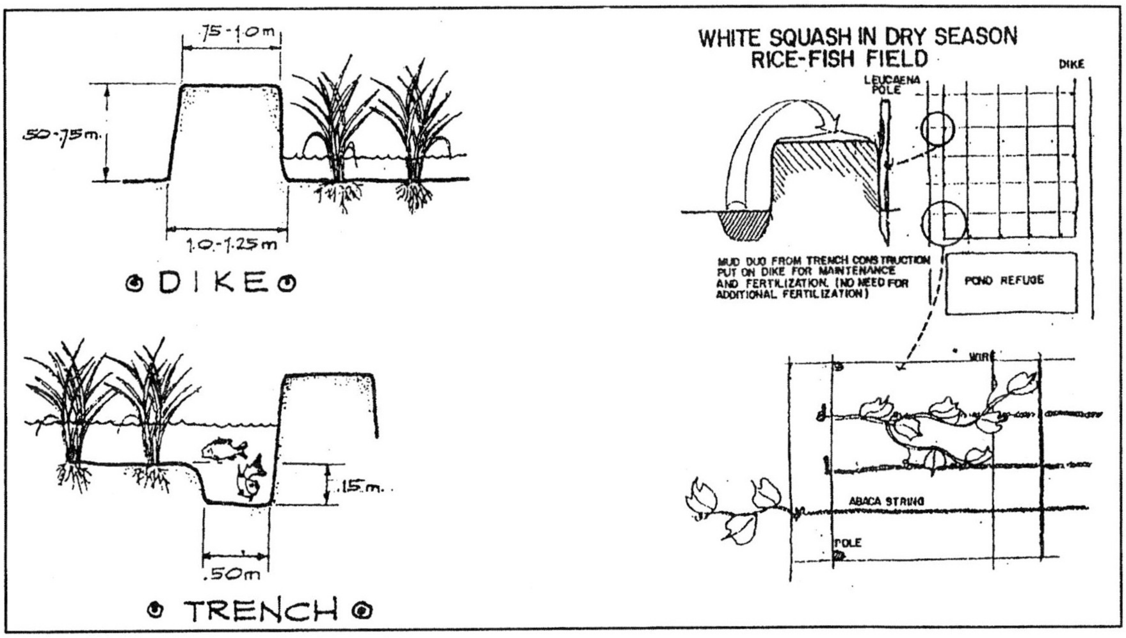

Farming systems

Mang Isko farms 2.3 ha of lowland with access to irrigation water from the National Irrigation Administration distribution system. Two rice crops are grown in 1.44 ha. Half a hectare is devoted to rice-fish culture. In some years, gourd is planted on the rice-fish dikes after the second rice harvest. Other vegetables occupy 0.14 ha of the farm where bittergourds are planted in the dry season and relayed with stringbeans in the wet season. The remaining 0.2 ha houses 1 pig in a 15 x 12 m shed and the rest of the area is grown to fruit and fodder trees and grasses.

Rice-fish subsystem

The 0.5 ha rice-fish system is composed of eight individual fields with side trenches. Two rice-fish plots have adjacent pond refuges in addition to the trenches. One rice-fish plot has an adjacent pond which is managed as a breeding pond. Mang Isko practices rice-fish culture in both wet and dry seasons and harvests two crops of rice and fish in a single year. However, when he plants gourd on the rice-fish dikes after the second rice-fish crop, he does not have a dry season rice-fish activity. In such occasions, the fields are drained and the fish are kept for growout in the pond refuges.

Combining fish with rice has doubled Mang Isko's rice yields in some cropping seasons. He attributes the increase in yield to these factors:

- Rice plants uprooted when digging the trenches are used to patch up vacant spaces in the ricefield where transplanted rice has not grown.
- The beneficial effect of fish on rice growth is manifested in the increased filleting of rice plants and the uprooting of young weeds when the fish (carp) stir up the bottom of the field in their search for food.
- The introduction of fish has meant that Mang Isko spends more time in his farm. Thus, he can spot and remedy problems immediately. In his own words, he has become "a better farm manager."
- Fish eat rice pests, thus rice yields are less threatened by pest damage.

Rice-fish culture practices

1. Land preparation, construction and maintenance

- Dike construction is labour-demanding. According to Mang Isko, it has been the biggest obstacle to rice-fish adoption. Collapse must be avoided and water seepage and overflow must be minimized in large dikes. They must be cleaned and weeded regularly to prevent damage by rodents.
- The trenches are dug one month after rice transplanting. The dugout mud is placed on the dikes for maintenance and is the source of fertile soil for the subsequent cultivation of gourd. Also, at this stage, the dugout soils are more compact as they have been soaked with water and this makes dike construction easier.
- Mang Isko uses one of the eight rice-fish fields with the highest elevation, as a test-field for monitoring water quality that comes in from the irrigation canal. This is to ensure that contaminated water due to pesticide applications of neighbouring farms do not get into his rice-fish fields. The irrigation water is let through this field first and any adverse effect on the fish is observed. The field is only lightly stocked (50 fish/800 m^2).

Mang Isko's monthly cash flows, showing cash expenses (cash outflows) and income earned (cash inflows) in one-year operation of vegetable production, rice monoculture and rice-fish culture

Months	January	February	March	April	May	June	July	August	September	October	November	December
Vegetables												
Cash inflow (Php)	bittergourd 1 000	bittergourd 3 200	bittergourd 1 400			stringbeans 600	stringbeans 1 000	stringbeans 2 600	stringbeans 1 000	stringbeans 600		
Cash outflow (Php)				stringbeans seed 120 urea 360	insecticide 280						bittergourd seeds 120 land preparation 80 urea 1 440 complete fertilizer 640 insectide 680	
Rice												
Cash inflow (Php)			rice 18 500							rice 30 500		
Cash outflow (Php)	urea 1 050 complete fertilizer 1 280		harvesting 2 000 threshing 2 000 drying 425 hauling 620 fish 1400		irrigation fee 670	land preparation 2 160	rice seeds 1 785.60 transplanting 1 440	urea 1 080 complete fertilizer 1280 insecticide 840		harvesting 3 200 threshing 3 200 drying 640 hauling 1 024 land preparation 2 160 fish 3 500	rice seeds 1 785.60	transplanting 1 440 insecticide 840 irrigation fee 575
Rice-Fish												
Cash inflow (Php)	fish 105	fish 70	rice 10 500		fish 1 420	fish 105	fish 140	fish 70	fish 70	rice 14 500	fish 1 400	fish 700
Cash outflow (Php)	urea 720 complete fertilizer 320		harvesting 1 200 threshing 1 200 drying 240 hauling 384		irrigation fee 690	land preparation 750	rice seeds 620 transplanting 500			harvesting 1 600 threshing 1 600 drying 320 hauling 512 land preparation 750	rice seeds 620	transplanting 500 irrigation fee 575

Initial cash costs for vegetable growing were incurred in the first months of operation

The cash requirements for rice monoculture and rice-fish culture were more spread over the months.

Unscheduled fish harvests in between two total harvests are a source of continuous income or cash inflows that help relieve cash constraints in certain months.

Mang Isko's calendar of farm activities

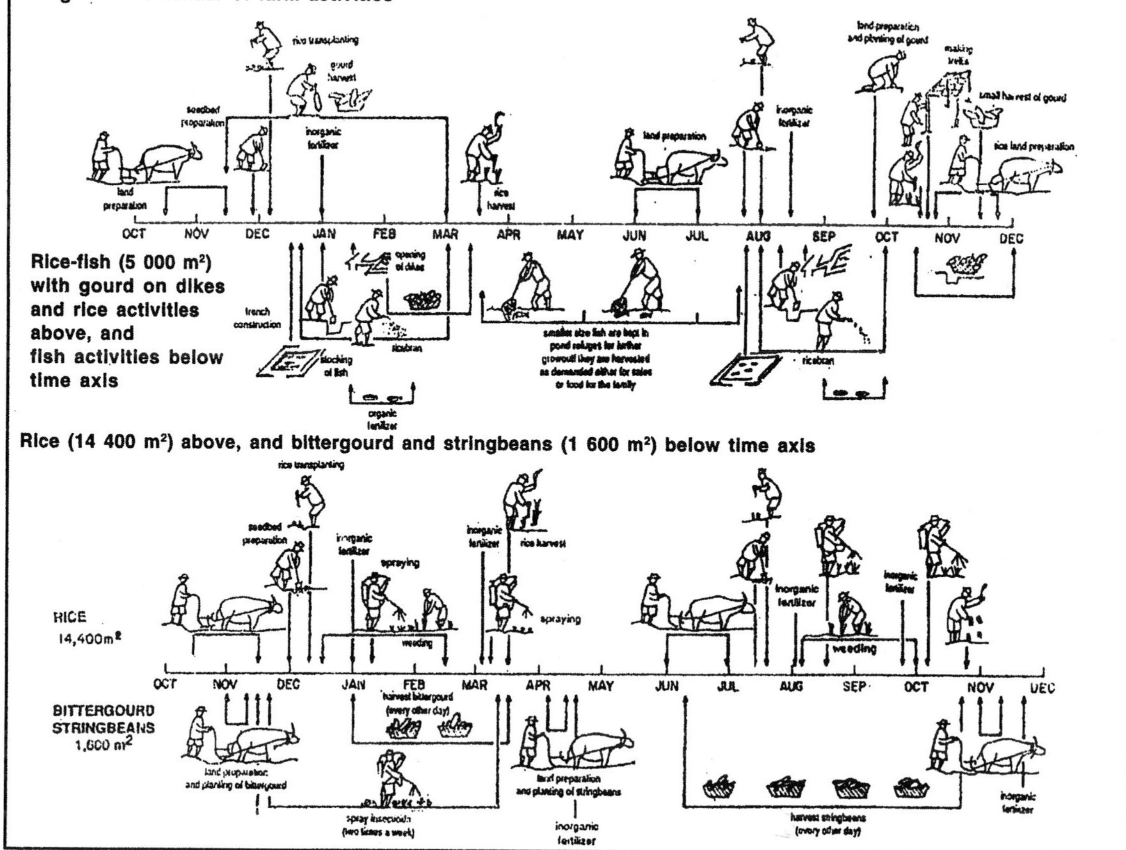

2. Rice transplanting and management
   - Rice is transplanted 10-12 days after sowing in seedbed.
   - Around one month after transplanting, three rows of rice, occupying about 60 cm, are removed for the purpose of trench construction. The uprooted rice plants are used to replace transplanted rice that has not grown.
   - According to Mang Isko, IR 64/74/42 varieties are not suitable for rice-fish as they easily lodge (i.e. stems bend and fold over).
   - He will try to use an early-maturing rice variety for the dry season to avoid any critical water shortage. At present, he uses a 90-day maturing variety for both seasons.
3. Fish stocking and management
   - Mang Isko keeps a separate breeding pond. Having one ensures fingerling supply. Moreover, he could stock large fingerlings which can be harvested as table-size fish immediately before rice harvest. However, without proper broodstock management, he had an inbred population after three years as reflected by stunted fish growth. After five years, he did not keep any broodstock.
   - Fingerlings are stocked in the pond refuges immediately after rice transplanting. After one week, the dikes connecting the pond refuge and the rice-fish plots are broken to let the fish in the rice-fish plots. (Stocking density: 1 tilapia/m² and 1 carp/5 m².)
   - The fingerlings/fish are graded into four classes when transferred from the breeding pond to refuge ponds and rice-fish fields: Class I, 25 pieces/kg; Class II, 35 pieces/kg; Class III, 40 pieces/kg; and Class IV, 50 pieces/kg. This is done in order to avoid cannibalism and

Mang Isko's monthly cash flows of all farm operations

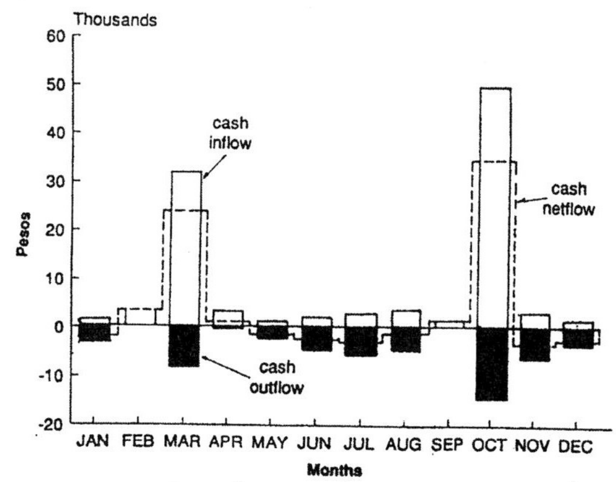

competition that would otherwise lead to large fish stunting the growth of small ones.

- One week after trench is constructed, it is filled with water and fish are stocked.

4. Fertilizing and feeding
   - Rice straws from the previous crop are burnt and the ashes are returned to the fields for liming.
   - Pig manure is thrown directly into the ponds or left by the inlet for water to carry/wash it into the rice-fish fields.
   - Fish feed on fallen rice flowers. Mang Isko believes that this has a purifying effect which counteracts the perceived off-flavour taste of tilapia due to the presence of pig manure in the system.
   - Rice bran is given one week after stocking in refuge ponds and trenches until two weeks before fish harvest. This is done twice a week or when required as may be signaled by the inactive behaviour of fish or its stunted growth.
   - Three weeks after transplanting, 100 kg urea and 50 kg complete fertilizer are applied to the ricefields.
5. Pest and disease management
   - Carp eat hatched golden snail eggs which drop into the water; tilapias feed on insects.
   - Mang Isko submerges the rice crop for 3 hours when insects become a problem. The fish then feed directly on insects on the plants as well as those trapped on the water surface. This practice is only carried out when the rice is 1-2 months old.
   - Mang Isko reports that a neighbour of his uses Gliricidia (*kakawate,* a nitrogen-fixing tree) as an insect repellant. In his first year of rice-fish, he placed Gliricidia branches approximately 1 m long at 2 m intervals around the edge of the field at the booting stage of rice, i.e. seven weeks after transplanting. He has now planted Gliricidia trees around the field as a means of biological pest control.
   - When constructing the dikes, a layer of plastic is placed on the inside of each dike. Rats find it slippery and difficult to penetrate the dikes with plastic lining.
6. Harvesting
   - The fish are harvested by draining the field 3 days before rice harvest. The water level in the refuge is lowered to several centimetres and the fish are caught by hand.
   - Table-size fish are sold. Fingerlings are kept for the next crop. Fish, which sizes are in-between, are returned to the pond refuge for further growout. They are consumed at home or sold, as a source of continuous income.

Monthly cash flows

- There were 5 months in the year where cash obtained from the sale of rice, fish and vegetables was greater than the cash spent on farm operations.
- The months in between rice and fish harvests were the period when cash deficit was greatest.
- Although there were sales received from vegetables before the rice and fish harvests, these were not enough to cover the large expenditures on inputs, particularly inorganic fertilizers.

Conclusion

As a whole, farming for Mang Isko was profitable. At the end of the year, he earned P45 233.80. He used this money primarily to sustain his wife and two children. A part of this was spent in upgrading his living condition, that is, he was able to improve his house and to purchase a refrigerator and a television set.

Issues for further consideration

The location of this single-household case study in Cavite, south of Manila, is classified as peri-urban, which has relevance on opportunities for sale of farm products. The area has undergone industrialization in the last decade and most of the farms have been bought up and the aggregated land converted to housing estates or factory installations.

Nevertheless, the case study illustrates the way the adoption of a rice-fish component has allowed further diversification on the farm. This is also based on labour availability and market opportunities for the new products. Part-time off-farm employment played an important role in household income for farmers in the area.

The farmer learned about integration opportunities from interactions with IIRR field staff. As most farmers are leaseholders, they need permission from their (usually absentee) landowners to modify the farm, i.e. dig deeper trenches and fish refuges, or even fishponds. This permission is often not granted.

Management for Rice-Fish Culture

Site selection: where to culture fish with rice?

John Sollows

Here are some factors to consider in selecting a site for culturing fish with rice:

1. Does the family have a particular area in mind? Whatever the answer, try to visit either the specific plot or the general area with one or more family members.
2. If the family already has an area in mind, ask what the members like about the area and take these into account in considering the following points.
3. Water (most important)

   The field must hold water continuously for several months; the longer, the better, for the fish. For best results, the field should be covered to a depth of about 30 cm, but if some areas are shallower or deeper than this, there is no serious problem.

   Does the farmer think he can achieve this? The higher-lying the field, the less water it is likely to catch. However, dikes and field boundaries must be above maximum flood level. The lower-lying the field, the more floodprone it becomes. At what level does the farmer feel sure he can control flooding?
4. Clay will hold water better than sand. Where does the farmer feel water will stand longest?

   If the field must be placed on a sandy area, generous manuring throughout the season will improve its water-holding capacity. How much manure can the farmer add?

   Form a compact ball from a handful of soil and drop it half a meter to your other hand. If the ball does not break, the soil holds water well. Successful culture is possible in poor soils, but faces more limitations.
5. How close to the farmer's house or "working shelter" can the field be placed? This makes checking the ricefield and feeding the fish less time-consuming. It also helps to discourage thieves.
6. Preparing the ricefield for fish culture is a lot of work. How can the farmer take advantage of existing conditions on his land to save effort? Some examples are given below:
   - A small knoll or termite nest can help provide part of the boundary for the field. This will reduce the length of the dike needed around the field.
   - If the land slopes, a high dike on the uphill side at the field is usually not needed. The layout of the land will help confine the fish.

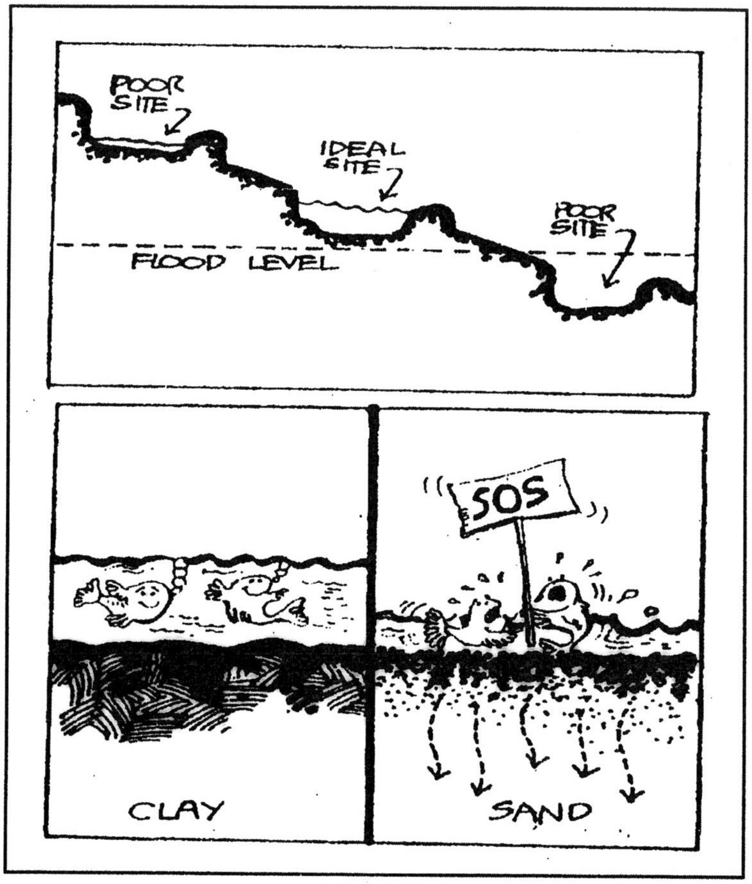

- Does the farmer have a pond within the ricefield already? If he can include the pond in his system, he may no longer need to dig a trench or pond.
- If the ricefield is basin-shaped, this can save a lot of work. The middle of the field is the deepest point and little effort should be needed to raise dikes. Is there any chance poisonous chemicals (industrial wastes, pesticides, etc.) will run into the field? Try to make sure this doesn't happen, since these poisons may kill all the fish.

8. The earlier a field is transplanted, the sooner it will be ready for fish. This means the fish may have a longer growing period.
9. The farmer may want to integrate his fish culture with his livestock, vegetable garden or other operations. In such a case, the site he selects may not be best for fish, but may be good for the whole operation.
10. Will the placement of the pond cause neighbouring fields any problem?
11. Any other considerations? Ask the farmer!

Issues for further consideration

It could be considered if fish should be introduced into direct-seeded (not described above) as opposed to a transplanted rice system. This is increasingly common with labour shortages in many parts of Asia.

The management of wild (i.e. unstocked) fish vs. stocked fish may have a bearing on system design and location. In many areas, e.g. in Bangladesh, there is a strong market demand for indigenous species. There the rural poor depend on small indigenous species caught from the seasonal floodplains, but these are becoming increasingly scarce as manifested in rising prices. Attempts at their cultivation, through entrance of spawn into the ponds from open waters, or through artificial reproduction and more intensive rearing in polyculture, which is being increasingly researched, may provide opportunities for smallholder farmers and their produce may fetch higher per-kilo prices and help meet market demand.

Designs may also consider the concept of the ricefield-to-pond continuum, with opportunities, advantages and disadvantages in cultivating stocked and unstocked species.

Preparation of field for rice-fish culture

John Sollows

Good preparation is very important in order to succeed in rice-fish culture. Every farmer must be able to:

- hold enough water over a large enough area for enough time to produce enough fish; and
- prevent serious flooding of the dikes and other boundaries of his ricefield.

Having a satisfactory water situation in the field is a key factor in the technology; this cannot be achieved if preparation is poor. In field preparation, there are four main things to consider: field size and shape, dikes, refuges and drains.

Field size and shape

1. How much land does the farmer own? If the farmer does not own the land and the landlord is agreeable, how big an area does the landlord want to try?
2. Topography and slope will greatly affect field size and shape. It may be possible to construct a large, square field on very flat land, but not quite so in sloping areas.
3. What area does the farmer think is suitable? This can limit field size and affect field shape. (See paper on site selection, this volume.)
4. How large an area does the family feel comfortable with trying out (especially for beginners)?
5. How large an area does the family think it can prepare and manage? (What does "manage" mean?) (See paper on feeding and maintenance, this volume.)

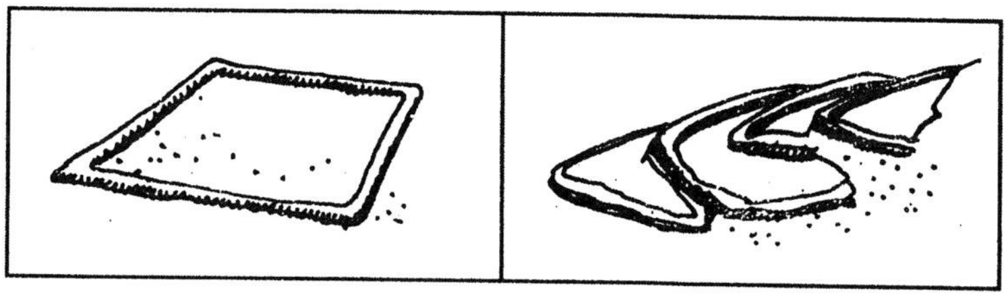

Some people say that a square field of 0.5-1 ha is the best size for rice fish culture. However, operations larger or smaller than this "ideal" size can also be very successful. Good preparation and good management are the keys to success, whatever the size.

Dikes

All dikes must be built safely higher than maximum flood levels. During construction, the dike should be raised high enough to allow for compaction and erosion.

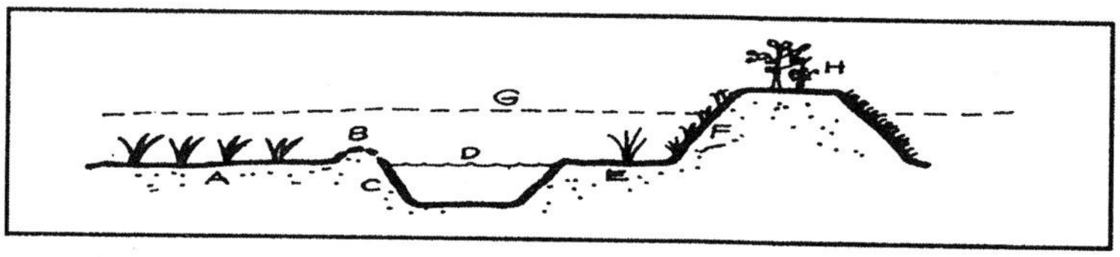

A - Existing ricefield

B - Small dike between field and pond/trench; optional but useful if fish are stocked in the pond/trench before rice is transplanted

C - Side of pond trench; slope should depend on nature of soil, more gentle in sand, can be steeper in clay

D - Level of water in pond/trench prior to transplanting

E - "Lip" of existing field between pond/trench and dike, to prevent eroded material from filling the dike; usually 0.5-1 m wide

F - Side of dike; slope depends on nature of soil (gentle in sand); with top soil and grass, erosion is reduced

G - Maximum flood level; point G is most important; the dikes should be high enough so they cannot be submerged by floodwater

H - Plants/trees grown on top of dike

I - Keep excavated topsoil and sods for the outside of the dike; compact the soil during construction, as possible

In raising the dikes, an excavation usually results. This may as well occur inside the field, all other factors being equal. This way, a small pond or trench is formed, which serves as a refuge for the fish.

Refuges

A refuge is a pond, trench or low point in the rice-fish field. When the rest of the field is dry, fish can be held here. Under some conditions (see paper on fry nursing in rice-fish systems, this volume), the refuge may be stocked before rice is transplanted.

Having a refuge is usually advisable and may be necessary for success. Without it, fish have to be harvested before the field dries out or moved to a pond in a flooded area. A refuge of at least 50 cm depth is desirable. If the farmer wants to hold fish all year around, the refuge will probably have to be much deeper than this.

In some well-irrigated areas, a refuge may not be necessary. Some farmers find that digging a refuge increases water loss. This can happen in cases where poor soil (like sand) is covered by the top soil, which seals water in. Digging a trench may break this seal; it will reform, but this will take time. Manuring speeds up the process.

A refuge, when dug, is usually made at the lowest part of the field so that water and fish can easily collect there.

Some other factors governing size and arrangement of refuges:

1. How much rice-growing area is the family willing to sacrifice for the refuge? This may depend on their total rice-growing area or on the relative importance they give to rice and fish.
2. How much money or time and labour can the family invests? As with field size, this can be an important limit.
3. What kind of soil is involved? A narrow trench (say 1 m wide x 1 m deep) will fill in quickly in sandy soil, but may last well in clay. The refuge in sandy soil should be three or more times wider than its depth.
4. Topography will affect trench or pond configuration. Extensive peripheral trenches on sloping

Some sample refuge layouts with comments

Plan / Cross-section	Comments
	Easy access for fish. Carrying excavated soil to dike can take time.
	Easy access for fish. Best for large fields on very flat land. Can be expensive to build. Difficult entry for buffaloes.
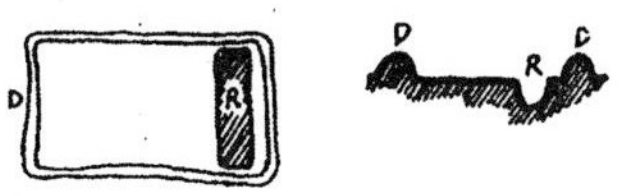	Widely applicable on flat or sloping land, especially for plots of less than 0.5 ha (but can work for larger plots, too).
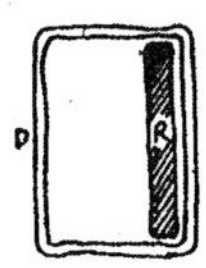  NOTE: D- DIKE, R- REFUGE, T- TRENCH	When trench is on low side of field on sloping land with porous soil, seepage can be a serious problem. Digging trench below ground level (rather than merely damming at ground level) and manuring can help. So can excavating the trench on the uphill side of the plot and sloping the plot toward the trench. This, however, can take a lot of work.
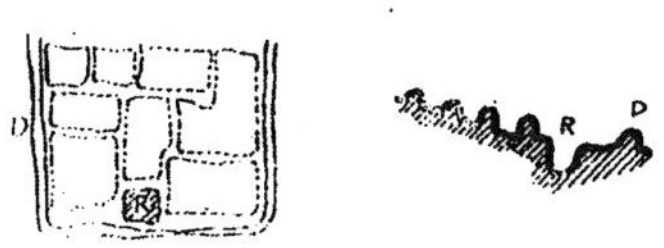	A common setup in rainfed northeast Thai systems. Small pond for refuge, in a system consisting of many plots on gently sloping land. Pond is usually at or near lowest part of field. Height of enclosing dike decreases as one goes uphill. This can help water catchment. Accommodates small-scale environmental variation, with little work. Farmers should be careful not to allow fish to have access to ponds when water is low.
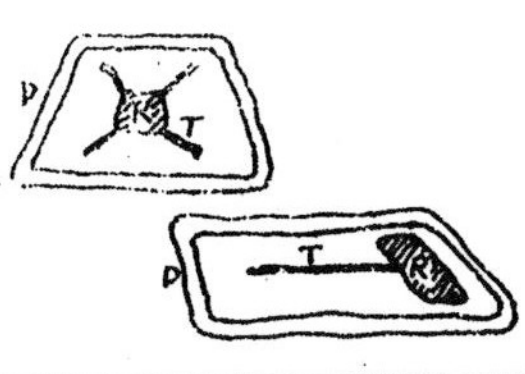	Narrow, shallow trenches connected to refuges can be very helpful to fish trying to reach the refuges. One or two rows of rice may have to be sacrificed.

areas will occupy too much space since such a field will be narrow.

Consider these two fields, each of 16 m² area:

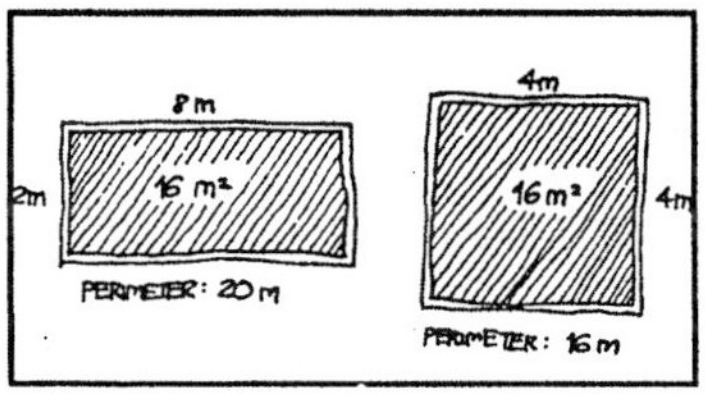

The narrow field has the greater perimeter to area ratio.

Drains

Usually, the field will need a drain so that excess water can be removed rapidly without eroding the dike. Inflow and especially outflow drains are advisable. Drains should be screened to prevent fish from escaping.

What material?

A bamboo, hollow log or pipe can be used, depending on availability. A screen should be placed at the point where the water enters. The screen can be a piece of fine netting or of flat metal full of nail holes. A little gravel scattered under the pipe will reduce dike erosion.

Such a drain is best for small fields (less than 1 000 m²) with limited flow (especially fields used to nurse hatchlings or small fry). Screens need to be checked every few hours for clogging any time water rises to pipe level and this can be a nuisance.

In most fields, the drain consists of a simple breach in the dike. This is screened by thin splints of bamboo or similar material, bound or nailed together.

Farmers in the rainier parts of northeast Thailand often use a *li*, a bamboo chute, set at a breach in the dike at the lowest part of the ricefield.

The *li* slopes up slightly and narrows. Below the narrow end, a jug-shaped basket or net bag is set. Hence, water runs out along the length at the *li*, but ultimately falls into a bag or basket which holds any fish washed out of the field. These can be eaten, sold or returned to the field for further growth.

Some farmers use a simpler version, by setting a net bag supported by sticks next to the outlet drain.

Outflow drain level

What depth of water is best for the rice in the field? What is the greatest depth it will tolerate? Set the drain in the dike somewhere between these two levels.

What drain capacity?

A small pipe does not drain a large field effectively. The farmer will have to make a guess as to how wide the drain should be, based on experience. It is better to have the drain a little too wide than too narrow.

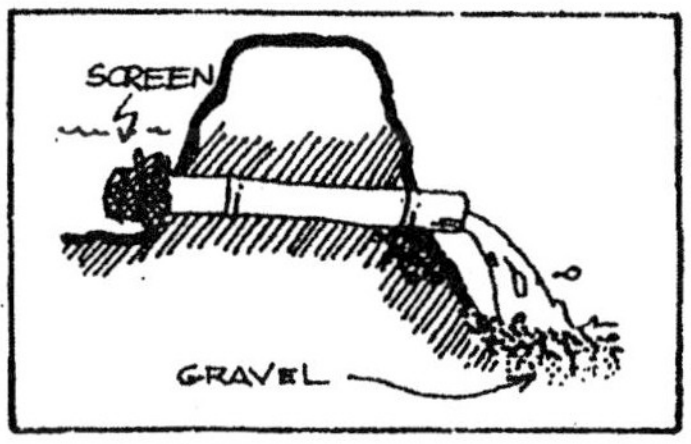

Stocking for rice-fish culture

John Solllows

The following guidelines apply in any case where seed fish are transported and stocked:

- Transport and stocking are best done early in the morning or failing this, late in the day when temperatures are lower.
- Fish, once purchased, should be transported promptly and kept out of direct sunlight.
- They should not be shaken up or unduly disturbed.
- On arrival at the pond or ricefield, bags should be set in the water (where the fish will be released) for several minutes until temperatures become the same inside and outside the bags.
- Bags should only then be opened and fish immediately allowed to swim into their new home of their own accord.

Timing

The earlier in the season that fish can be stocked, the longer the growing period. Also, the earlier in the rainy season, the fewer the predators.

However, fish cannot be stocked before there is water available and the farmer should be reasonably sure that the field will hold water for several months before he stocks. Rice should also be well-established with 2-3 tillers out before fingerlings or large fish are allowed into the field.

Finally, the farmer may be ready to stock but seedfish may not be available. Therefore, the family may have to wait until fish can be found.

Stocking rates, sites and species

There is no "best" formula here. Large fish are more expensive than small ones, but are better able to escape predation. Species can differ in price. Many fish cost more than few; the family's budget, then, can affect what is stocked.

For beginning farmers and for those who cannot feed their fish, stocking not more than 300 (5 cm) fish per 1 000 m^2 is suggested. A species ratio that commonly works in Thailand is common carp *(Cyprinus carpio)*, silver barb *(Barbodes gonionotus)* and tilapia *(Oreochromis niloticus)* at 2:2:1.

This formula will not be appropriate in every case, but is as good a point of beginning as any. In general, it is better to culture two or more species than only one, since different kinds of fish eat

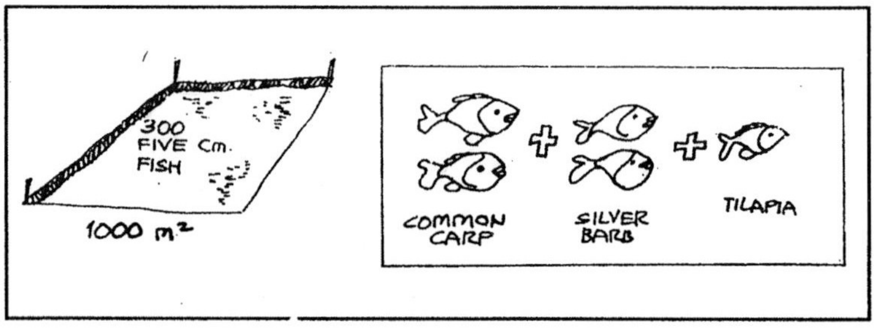

different foods. This means total catch should be higher than if only one species is raised. The formula given can be modified for many reasons:

1. Availability

   A farmer may want a certain combination of species and sizes, but still has to be content with what he can find on the market.

2. Preference

   Each family will have different species preferences, usually for valid reasons. These should be accepted. Similarly, many farmers prefer large seed because of their higher survival or greater final size. Others prefer small seed despite probable higher mortalities because of lower prices and higher continuity of harvesting; specific prices will affect the economics here. The family with limited budget must often decide between buying a few large fish or many small ones.

3. Species-specific biology

   Fish species have differing advantages and disadvantages: Tilapia *(O. niloticus)* tolerates environmental extremes very well and reproduces easily. The farmers who can keep a few fish all year round need not worry about restocking every year. However, reproduction can lead to overcrowding and poor growth. Some farmers do not like the taste, find the fish ferments poorly and complain that it competes with or drives away other desirable species.

   Common carp *(C. carpio)* tolerates poor water quality and shows excellent growth in most ricefields. However, probably due to high susceptibility to predation, survival of this fish is poor.

   Silver barb *(B. gonionotus)* usually has excellent survival in ricefields; even fry tend to show good recovery. It is less tolerant of poor water quality than the two species above and does not grow well in very shallow water or one of highly unstable depth.

   Various wild species, notably snakehead (*Channa* sp.) and walking catfish *(Clarias* sp.) are very palatable.

   The snakeskin gourami *(Trichogaster pectoralis)* has shown very promising results in a few rainfed ricefields. Broodfish, not seed, should be stocked. More work should be done on this species under rainfed conditions.

   Chinese and Indian major carp usually show poor growth in rainfed fields. In deeper water (50 cm or higher), they appear to do better. They should be stocked at low rates, no more than 200/ha.

4. The presence of several important predators can affect size and species stocked. Large fish escape predators easily, but this appears a less important consideration for silver barb than for other cultured species.

5. Culture field will often affect number and species stocked. Occasionally, silver barb will not grow well in field with very shallow water (less than 10 cm). In small fields, the farmer may

If stocking density is low, there is often sufficient natural food in the paddy and no feeding is necessary. **If stocking density is increased, natural food in the paddy is not enough and production is low.** **If stocking density is increased, maximum production can still be obtained with supplementary feeding.**

find the advisable number of fish limited by available area. On the other hand, there is nothing wrong with stocking few fish in a very large field, especially if this is all the farmer can afford.

6. The suggested rate of 3 000/ha can be increased if the field has stable water depth (30 cm or more is preferable) and if the field can be fertilized frequently. If fish are fed, the feed should be put in the field, not in the refuge. Otherwise, they will stay in the refuge, the rice will not benefit and the fish will become overcrowded. Farmers should be very cautious about stocking over 6 000/ha. This can work occasionally, but should be done only by experienced farmers who know their system. Small fry can be stocked in greater numbers than large fingerlings.

Issues for further consideration

The case presented here depicts fingerling transport in oxygenated plastic bags, which are not widely used in poor rural areas, for example in Bangladesh and India. There, other delivery systems, such as in open containers, are used.

The stocking rates and practices shown in the example imply that food fish are to be produced. On the other hand, ricefields can be used as nurseries for fry, the practice of which is well tuned to the short-growing duration of many rice varieties (e.g. 3-4 months). In such cases, the stocking densities would be considerably higher than the 3 000 fry/ha suggested above.

Depending on local availability, other species combinations would be chosen. The roles of different species more suitable for ricefields should be considered. Surface feeders (e.g. silver barb) and bottom feeders (e.g. common carp) perform relatively better than column feeders (e.g. silver carp) in ricefields as compared to ponds.

As a simple prophylactic measure that helps against infectious fish diseases, fish seed could be given a dip treatment for 1 minute in a 50 percent common salt solution, prepared on the spot at the site of stocking.

Feeding and maintenance in rice-fish system

John Sollows

Maintenance check

Daily check the water level in the field to see that it is not rising or falling unusually quickly. If this occurs, find out what is causing it. Any leaks should be clogged. A shovel or hoe should be carried on these visits.

Some farmers throw a little feed every day in order to monitor their fish stocks.

In intensive systems, early morning checks to see if fish are gaping are advisable.

Feeding and fertilizing

Feeding and fertilizing should normally help fish grow. However, it is not a major consideration in lightly stocked fields (below 3 000/ha), where fish should be able to forage sufficiently for themselves.

Families who would like to stock more heavily (and therefore to feed and fertilize) need to consider the following:

a. Will they have time to feed or fertilize well? (How far away is the field from their house? What other work do they have to do?)

b. Can they get feed or fertilizer? Is it easily available in the area? Is it affordable?

Types of feed and fertilizer

It is difficult to draw a line between "feed" and "fertilizer" especially since manure can be used as both. Inorganic fertilizers can be used. So can any nontoxic organic material.

Manure is often the most important addition, by weight. Either fresh or dried manure can be used. A little caution with fresh manure may be needed if water is stagnant, but it has been observed that up to 300 kg/ha/week go into such systems without causing harm. Replenishing manure as the fish consume it is another way to cope.

Rice bran is commonly used as a fish feed. It works well in nurseries, but is usually not needed in extensive rice-fish culture. If farmers have to pay for it, they probably should not use much, once fish have entered the field.

Some farmers use rice hulls in their systems and some fish species eat these eagerly. Most of the hull is not digested, but gets spread around the field by the fish.

Kitchen wastes and leftovers of any kind can also be given.

Different kinds of water plants work well: *Azolla, Wolffia,* duckweed *(Lemna), pak boong* or kangkong *(Ipomoea aquatica)* and water mimosa are examples. Different fish species will have different preferences but silver barb will eat any of these.

Crop by-products are also acceptable: cabbage leaves and corn cobs have been used by some farmers. Cassava leaves are also popular. Since some cassava varieties may be poisonous, it is advisable to dry cassava leaves before feeding them to fish.

Termites are a very nutritious feed and are especially helpful in nurseries. Nests are chipped over the pond or field and the termites fall into the water, where they are rapidly consumed. Termites are usually not needed once fish have entered the ricefield; if farmers continue to use them heavily throughout the season, they may run out of nests! Other insects, shrimps and worms are similarly nutritious.

> Caution: Termites have been reported to cause fish mortality if fed live as these are swallowed whole. The termites continue to bite with their large jaws while in the stomach, causing fatal injuries and perforations to the stomach wall. A safe way to utilize this excellent food source is to drown the termites in a separate pan or bucket before feeding them to the fish.

Rice straw is not usually eaten directly by fish, but feeds small plants and animals on which fish feed. It can be used anywhere, but may be especially helpful in turbid nursery ponds.

Any otherwise unused dead animals, entrails or body parts can be put to use. In ricefields, they can go directly into the water for fish consumption. In nursery ponds, large, decaying animals can contaminate the pond. Some farmers suspend animal parts over the pond. These attract flies, which lay eggs on the meat; maggots can then be knocked off the meat into the water to feed the fish.

Jute or *kenaf retting* can make water temporarily unsuitable for fish culture. This is a process in which the freshly cut plant stems are submerged in ponds and ditches to let the soft plant material rot off, with the desired fibers remaining, which are dried and processed. In these ponds, the water turns black, oxygen levels drop to near zero and the water smells bad. The *retting* is very effective, however, in clearing up turbid water. After the *retting* is finished, pond water quality is often improved. Also, small amounts of *jute* or *kenaf* will not harm fish and the rotting material provides feed. Larger amounts can be placed in stagnant water. Good figures for safe rates for fish, unfortunately, are not available, so only small amounts should be used and the fish should be checked every morning to see if they are gaping.

Other examples of feeds include mulberry leaves, banana leaves, bat or livestock dung, animal feed leftovers, coconut oil residues and *Leucaena* leaves. No list of potential feed stuffs will be complete.

Management

In densely stocked fields (over 5 000/ha), continuous feeding and fertilizing become important, particularly as the fish grow. Giving small amounts of feed a couple of times a day may be advisable. Check to see how quickly a known amount of vegetation or manure gets consumed. If some amount remains after an hour, there is no need to increase the rate. If it disappears within half an hour, increasing the amount is advisable.

> **Issues for further consideration**
>
> The example relates to very extensive rice-fish growout systems with low stocking rates as implemented in northeast Thailand, which do not require feeds and fertilizers. On the other hand, more intensive rice-fish systems for nursing or food fish as they exist elsewhere in Thailand and other countries have higher stocking densities, frequent water exchange and, most importantly, specific feeding and fertilizing methods.

Rice management in rice-fish culture

John Sollows and Catalino Dela Cruz

Rice-fish culture can be carried out under rainfed or irrigated conditions, in either direct seeded or transplanted fields. Timing of seeding and transplanting activities are affected by many factors (water availability, rice variety, etc.) but are not usually affected by the fish culture component.

Seedlings are best transplanted 25-30 days after seeding although the best age for traditional varieties may fall outside this. In practice, they often remain in the seedbed longer than this. Sometimes, droughts occur so that the fields are too dry to be transplanted and the farmer must wait for rain. In other cases, the family labour force is limited and the rice in the seedbed must "wait" until the family gets to it.

Most farmers find no problem in applying chemical fertilizers to their rice-fish systems. In some cases, it has been reported that fish die after exposure and when they are fed with pellets and may have ingested fertilizer granules for this reason.

The wide scale of rice-fish is still constrained by continued application of pesticides in rice-based farming. The use of pesticide is not recommended in rice-fish farming. In rice-fish culture, there are ways of controlling rice pests that do not need pesticide, such as:

- Quick submergence (for three hours) of rice plants in water. This makes the insects vulnerable to fish predation. Limitation: suitable only before plants are taller than the dikes.
- Two persons can drag a stretched rope (50-100 m) across the rice fields to knock off the insects into floodwater, after which they can be eaten by the fish. Limitation: suitable only before rice plants reach booting stage.

However, should a farmer insist on using pesticides, here are some helpful tips:

1. Considerations in applying pesticides:
   - Choose and apply properly pesticides that have low toxicity to fish.
   - Minimize the amount of pesticide getting mixed with water.
   - Apply at suitable time.
2. Considerations in preventing fish poisoning:
   - Drive the fish into the sump, draining the field slowly before spraying. Keep the fish in the sump until the toxicity in the sprayed field is gone.
   - Increase water depth (+10 cm) to dilute the concentration of pesticides in the water.
   - Flush water through the ricefield. Open the inlet and outlet of the field, and allow irrigation water to flow freely during spraying. Begin spraying from the outlet end of the field. When one-half of the field is already sprayed, stop for a while and allow the pesticides to flow out of the field. Then, continue spraying towards the inlet end of the field until it is finished.

To do second and third items above, here are some examples: apply powder pesticides in the morning when dewdrops are still on the leaves; and apply liquid pesticides in the afternoon when leaves are dry.

There are a number of less toxic pesticides in the markets. Application of a toxic insecticide like Furadan or Curaterr ® can be made safe to fish if done properly through solid incorporation during the final harrowing. Furadan is a systems insecticide the efficiency of which in controlling insect pests lasts about 50-55 days. Incidence of pests after this period can be controlled by spraying liquid pesticides. At this time, the rice plants are already in their full vegetative stage and the thick leaves will intercept most of liquid sprays, thus drastically reducing the concentration of pesticides reaching the water.

It is best to wait until the rice is well established before releasing seed fish, particularly if the fish are large. Fish can be stocked once two or three tillers have appeared for which the usual waiting period is 1-3 weeks after transplanting or 4-6 weeks after direct seeding depending on the state of the rice and the size of the fish.

Small fry (about 2.5 cm long) can be stocked immediately after transplanting, without harm to the rice. The authors have never seen a rice variety that does not work with fish, but some varieties are better than others. Deepwater-tolerant varieties are preferable to those which thrive in only very shallow water.

In some areas where rainfalls

are highly unpredictable, farmers prefer to wait until very late in the rainy season to stock fish. At this time, surface water accumulation will be at its yearly peak and the chance of flooding from later rains is very slim. In such cases, long-lived, late-maturing rice varieties are best.

Rice varieties which tiller (i.e. produce new plant stems) rapidly or under a wide range of water conditions will allow farmers to stock earlier in many cases.

Farmers have succeeded with early and late-maturing photoperiod-sensitive and nonsensitive, glutinous and nonglutinous varieties.

Effects on rice yield

The authors' experience indicates that rice yields rise on the average, by about 10 percent, in rice-fish situations. However, there is great variation from farm to farm so guarantees cannot be made.

Yields seem the most enhanced on farms with poor soil where fish are fed intensively. Possible mechanisms include:

- increasing availability of nutrients for increased floodwater productivity and uptake by rice; and
- reducing loss of ammonia through volatilization after fertilizer application by preventing floodwater pH to rise over 8.5.

The greatest danger to rice has already been indicated: big fish will damage very young rice; otherwise, some rice varieties do not tolerate deepwater. By using very sensible precautions farmers are not likely to harm their rice yields.

Issues for further consideration

Today, integrated pest management (IPM) is the declared national pest management strategy in the Philippines and many other rice-producing countries, and the International Rice Research Institute has published results that natural control of rice pests without pesticide use generally is the most profitable option for rice farmers. The concept of IPM certainly precludes the use of systemic insecticides for preventive treatment.

Rice-fish benefits and problems

John Sollows

In discussing a technology with potential new entrants, it is important to acquire it with potential benefits and risks so that they can make as balanced a decision as possible as to whether or not to try out the technology. If the potential new entrants are not aware of the possible benefits, they may miss a chance to improve their standard of living. Ignorance of the risks can also lead to serious problems and reduce the new entrants' self-reliance.

Problems and limitations

1. Rice-fish culture requires land. Landless farmers will have difficulties here unless they can make arrangements with the owner which must be mutually beneficial. Acquainting the owner with the benefits and problems associated with the technology will be important. The agreement should spell out what part of production goes to the farmer and what goes to the owner. Will rent be increased? Will it be rearranged? Will all additional benefits from fish catches go to the farmer?

   Can rice-fish culture on communal land be arranged for landless farmers?
2. Production cannot be guaranteed especially in rainfed situations.
   - Good water management is essential but not always possible. Rain is not predictable nor controllable. Too much water can lead to flooding and escapes. Too little water inhibits growth and, in extreme cases, can kill fish. Fish cannot be cultured without water. (See papers on preparation of field and on feeding and maintenance, this volume.)
   - Poor water quality can impede growth and cause death. This is rarely a problem in ricefields, but can be an issue in nurseries. (See paper on fry nursing in rice-fish systems, this volume.)
3. Pesticides and other toxic chemicals can kill fish and should be kept away from them. (See papers on site selection and on rice-fish culture, this volume.)
4. Transport of seed fish and stocking should be carried out correctly. Seed fish are very vulnerable at these stages; carelessness can kill. (See paper on stocking for rice-fish culture, this volume.)
5. Predators can seriously reduce fish stocks. Food nursing can solve this problem to a large extent. Submerged snake traps of wire mesh can be used to drown snakes. Snakes and frogs can also be caught manually. Frog eggs should be removed when discovered and dried. Birds can sometimes be scared away.
6. Thieves are perhaps the most difficult predator to deter. Living near the field helps sometimes. Partly filling the pond with bamboo or other branches makes netting the fish difficult and submerged barbed wire will probably ruin any net it snags. Obstacles (rocks or logs) placed on the dikes leading to the field makes access difficult at night. Watchdogs can also help.

Too much water

Too little water

7. Field preparation will demand a large investment of time, labour and money from the family. For poor farmers, labour availability often affects their ability to carry out the practice, limits the area they can prepare and affects the intensity with which the system can be managed. Old and young couples with small children will

be particularly challenged here. As a rule of thumb, a 1 000 m^2 field will not take more than ten 8-hour days to construct, if one person is doing the digging. A family with no time to feed the fish should stock lightly.

8. The farmer's managerial skills will increase in time. Many farmers succeed during first year, but many fail, as well. Failure among experienced farmers, however, is seldom
9. Rice yields are occasionally reduced by rice-fish culture. This occurs most often when large fingerlings are stocked before the rice is well-established. The water in some fields as well may be deeper than desirable for some rice varieties. Sometimes, rice will lodge and fish will graze on the rice seeds.
10. Some farmers complain that wild fish catches are reduced in fields with cultured fish. Tilapia is most often indicated as the suspect. These farmers feel that cultured fish in large numbers can scare away wild species.
11. Marketing problems can occur. A farmer can plan to keep his fish to sell when prices are high, but water shortage can force him to sell before this. Transporting fish to the market can also take time, especially when arrangements cannot be made beforehand. If a family plans to sell an important proportion of their catch: where, when and how will they sell it? Will this be easy?
12. Seed fish supply is a very common problem. A family may not always be able to get what it wants. Seed fish purchase usually occurs during the transplanting season when demands for fish are high and farmers have little time and money.

In villages where fish culture becomes widespread, the establishment of small hatcheries and nurseries deserves serious consideration. It is often advisable to encourage two or more interested villagers who feel that they are in a position to manage such operations to start the hatcheries, if the demand on the local market is sufficient. This will keep one producer from monopolizing the market.

Benefits and potentials

1. Compared to many technologies, rice-fish culture is a low-risk technology. It demands little money, is not particularly "new" or revolutionary for most rice farmers and involves few conflicts with other farm activities.
2. Fish cultured in ricefields provide farmers with a continuous, predictable, convenient supply of food. Farmers accustomed to depending on uncertain, declining stocks of wild fish appreciate this.
3. Rice-fish culture conserves water.
4. Rice-fish culture saves farmers time, allowing them to undertake income-generating activities or to improve on existing ones.
5. The small amounts of money needed mean that farmers need not take out loans. They, therefore, have many options as to how to use their fish: eat them, sell them, keep them alive (nature permitting), preserve them or give them away. The farmers do not have to make quick sales to reduce debts.
6. Income from sales can provide useful money at various times. Some farmers can sell brood fish or seed fish, as well as table fish.
7. Since this is a subsistence activity, to a large extent, there is little competition on the market among producers.
8. Rice yields are usually enhanced, although there is great variation from farm to farm. Yields are very rarely adversely affected when the farmer manages the system well.
9. The fact that this technology can modestly improve the lives of many poor rice farmers should make it of interest to development workers.

Issues for further consideration

One of the major constraints of the rice-fish system can be the submergence of the paddy fields due to seasonal floods, which leads to either loss of fish or a mixing of these with those from neighbours, which are then found in the ricefield once the flood pulse recedes.

Studies have shown that the existence of tilapias can increase the biomass of harvestable snakeheads in ponds. Changes in water flows and access to ricefields after modification of fish culture may be more of a problem.

The role of perennial water availability for continuous fish holding capability, i.e. a deeper sump or pond, may be considered. This may be valuable given marketing problems and for improving the availability of fish. Ponds allow more flexible marketing. While fish raised in ricefields may only be available for a short period, the option of storing them in a pond extends the fish availability season. Water conservation in rice-fish culture requires a deeper pond to be part of the rice-fish system.

In areas where per capita fish consumption is high, the cash saved from having to purchase fish can be a strong incentive to raise fish in ricefields. In irrigated areas, a high-input rice-fish system can improve subsequent crop yields and/or reduce nutrient requirements.

Economics of the system can vary. In northern Viet Nam, for example, income from the rice-fish system is often 1.5 to 1.7 times higher than from the rice-only system. While rice productivity in the rice-fish system is 10-17 percent higher compared with the rice-only system, the total rice production is often only 3-5 percent higher when considering the rice cultivation area lost to the trench. Another benefit of the system as experienced in northern Vietnam is the 50-65 percent reduced use of pesticides compared to the rice-only system.

The rice-fish ecosystem

Ahyaudin Ali

Figure 1. Simplified nutrient flows

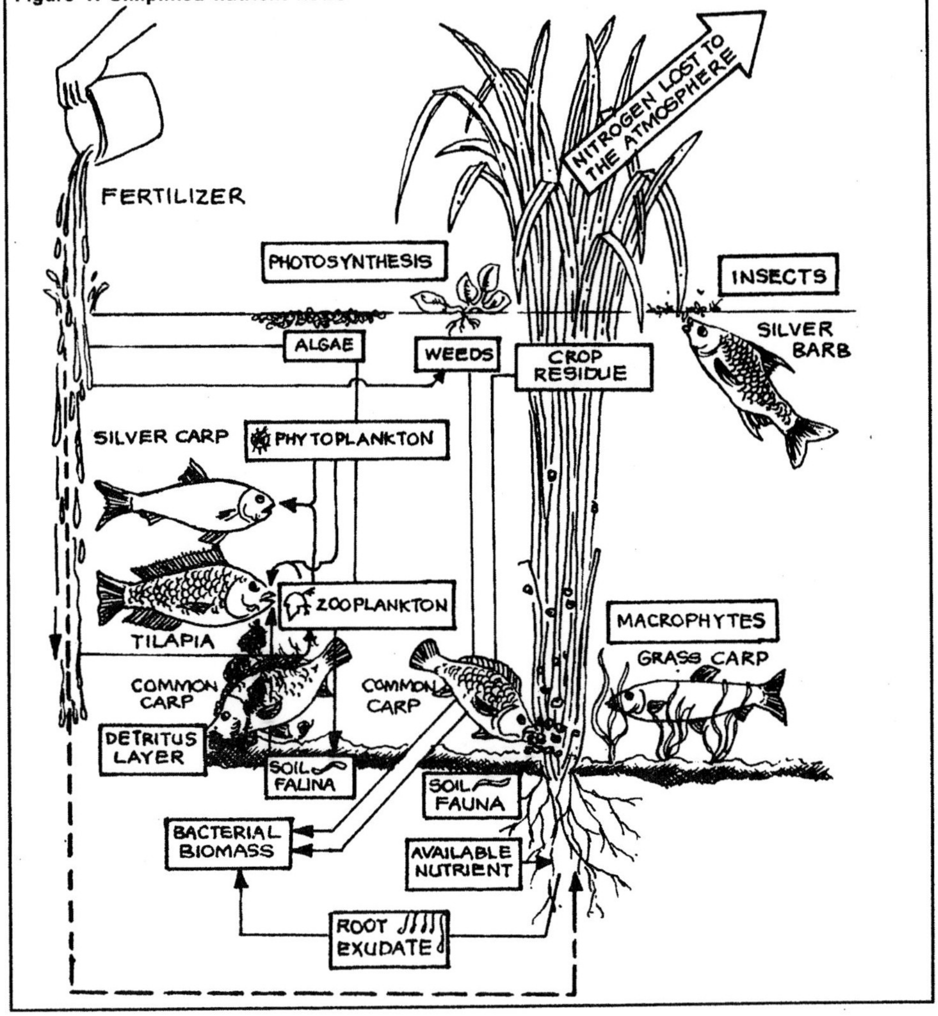

To understand the rice-fish ecosystem as discussed in this paper, here are some terms to note:

- Fertilizer - input of simple nutrients to the system.
  - organic - available also to fish, plankton, algae, soil, fauna and bacteria
  - inorganic - available only to rice crops, macrophytes/algae/weeds, phytoplanktons and bacteria.
- Photosynthesis
  - produces food from simple nutrients using sun's energy;
  - plants, algae and phytoplankton are food for fish, insects, zooplankton and soil fauna
- Decomposed materials - add to the detritus layer
- Bacteria - organisms which recycle materials back to simple nutrients
- Fish - examples of various feeding groups

Figure 2. Ecosystem components

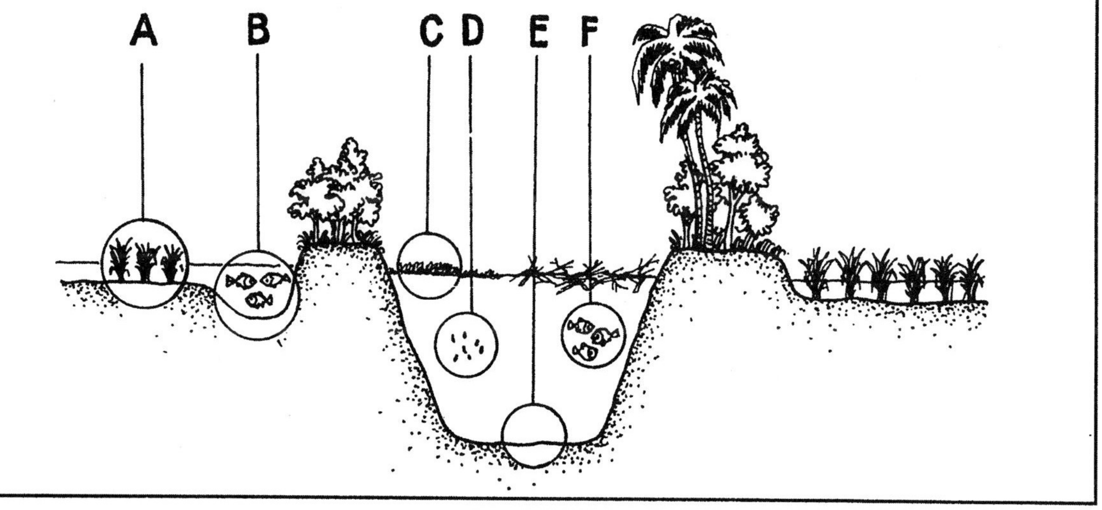

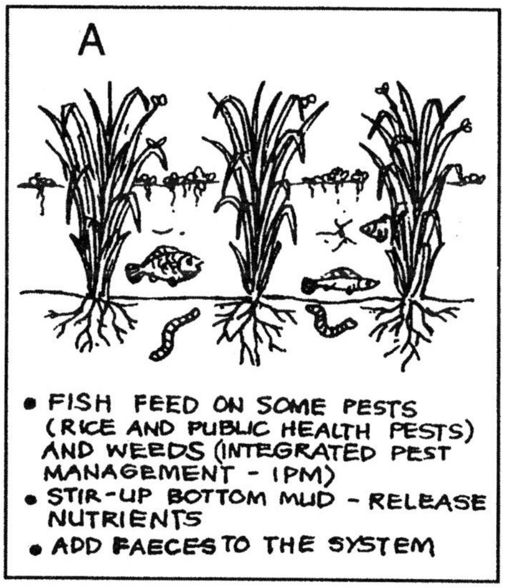

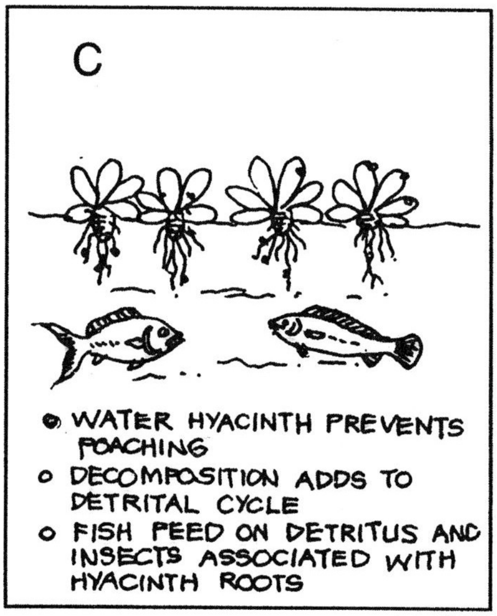

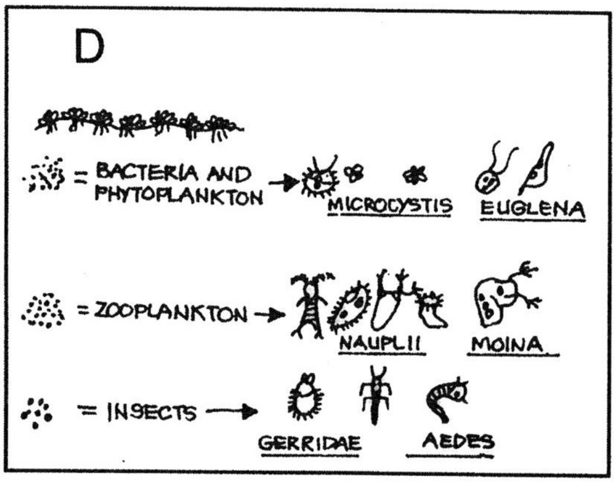

E

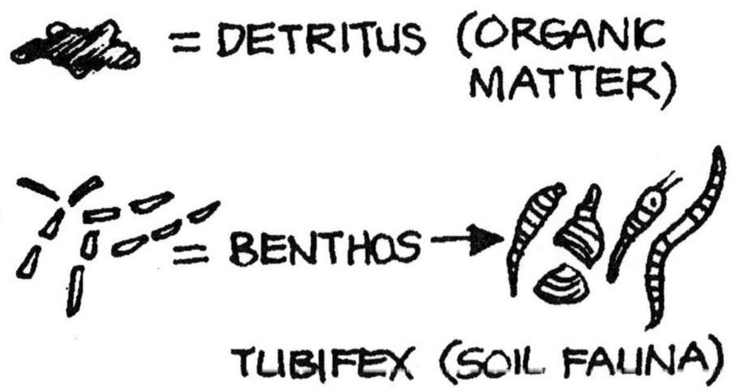

F

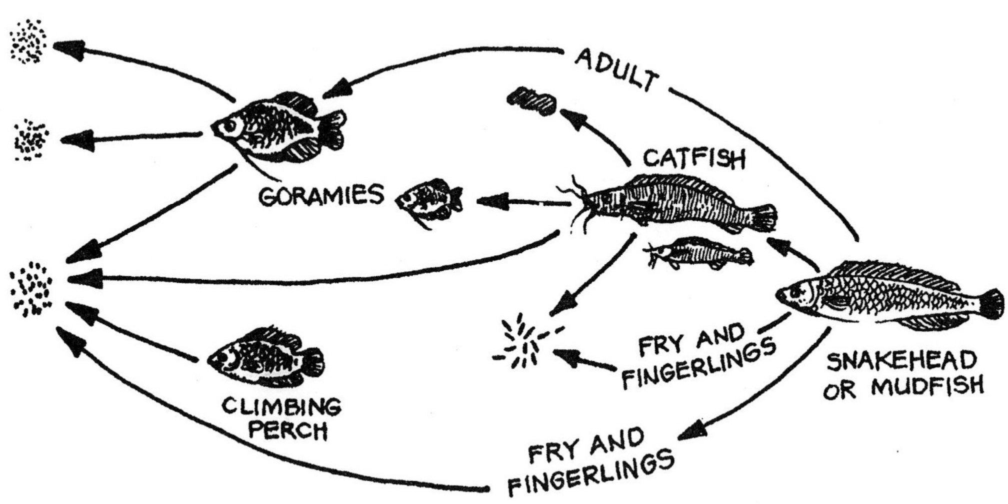

Conceptual food chain for fish in the system

Issues for further consideration

The ricefield environment is a specialized culture environment for fish that can be exploited best through certain types of polyculture.

Fish such as common carp, through their foraging behaviour, aerate the topmost sediment layer and enhance the aerobic bacterial decomposition process thereby ensuring the release of inorganic nutrients at a rapid rate.

Fish as a component of integrated pest management in rice production.

Matthias Halwart

Integrated pest management

Integrated pest management (IPM) is a pest control concept that uses the best available mix of technologies for a particular pest problem. It promotes practices available to the farmer and permits pest control with the least use of chemicals to get high yields and maximum profits. Reducing pesticide application is an important factor because this control method is often uneconomical and unsafe to humans. Moreover, many pesticides kill both pests and their natural enemies at the same time, leaving the crop open to an uncontested invasion of pests.

IPM includes a variety of technologies, such as choosing pest and disease-resistant varieties, using crop rotation, fallows and simultaneous planting over wide areas. It also encourages the establishment of natural enemies of rice pests.

Economic threshold levels are used to help farmers decide when the application of pesticides is economically justified. The presence of parasites and predators, including fish, minimizes the need for the farmer to take action.

Fish as predators in ricefields

The most widely spread fish species in paddies are common carp (*Cyprinus carpio*), Nile tilapia (*Oreochromis niloticus*) and silver barb (*Barbodes gonionotus*). However, there is a large additional number of both stocked and wild fish species in ricefields.

Among those, there are larva-feeding and mollusc-feeding fish which are of considerable importance in the control of human vector-borne diseases, like malaria and schistosomiasis. Plant-eating fish species directly feed on weeds and are reported to be efficient in keeping irrigation canal free from vegetation. Turbidity and high water level add to the effects of fish controlling weeds.

Different fish species will affect different pests in different ways. Common carp, for instance, appear very effective at controlling golden apple snail. In Malaysia, tilapia was not able to control this pest but the giant walking catfish performed well.

The effectiveness of fish in fields where rice is broadcast is doubtful, since the rice will be too crowded to give fish good access to the field. There is no evidence for an optimal fish stocking density for pest control, but higher densities are presumed to be more effective. Fish eat older, outer rice leaves which are more likely to be infested with pathogens. This will make the rice plants healthier.

There are several reports documenting either reduced num-

Table 1. Problematic organisms in rice production, predation of fish and susceptible life stage of prey.

Problematic organisms	Predation by fish	Life stage of prey
Insects		
Whorl maggot	?	?
Gall midge	+	?
Caseworm	+	Larvae
Leaffolder	?	?
Stemborer	+	Larvae
Black bug	?	?
Brown planthopper	+	Nymphs/adults
Whitebacked planthopper	+	Nymphs/adults
Green leafhopper	?	?
Diseases		
Rice blast	?	?
Sheath blight	+	Sclerotia
Bakanae	?	?
Brown spot	+	?
Bacterial blight	+	?
Bacterial leaf streak	?	?
Weeds		
Grasses	+	Seedlings
Sedges	+	Seedlings
Broad-leaved weeds	+	Seedlings
Others		
Golden apple snail	+	Hatchlings

Notes:
1. ? - no information available
2. + - based on documented observations

ber of pests or less damage caused by pests and diseases with the presence of fish (Table 1). Most of this work has been done in concurrent rice-fish systems. In some cases, the underlying mechanisms have been described (Figures below). However, this list is far from complete. It is likely that more direct and indirect effects will be revealed with continuing research.

To date, feeding habits of fish in the ricefield habitat are not well-established. Generally, younger fish are more omnivorous and insects often form part of their diet. Abundance and behaviour of fish and rice pests will determine the relative importance of fish as a biocontrol agent in rice production.

Field observations are the basis of understanding the role of fish in ricefield ecosystems. If you are working in the field, please record and share your experiences with others on the activity pattern and feeding habits of fish. The important factors are species, size, prey and response to environmental changes.

Pest control mechanisms

Golden apple snails

Fish feeding on newly hatched snails

Stemborer

Fish feeding on dispersing stemborer larvae

Planthopper

Fish feeding on hoppers falling on water surface may actively shake rice hulls when nibbling on the stems

Caseworm

Fish feeding on floating case-worm larvae

Rice sheath blight

Fish feeding on floating sclerotia

Direct/Indirect

Weeds

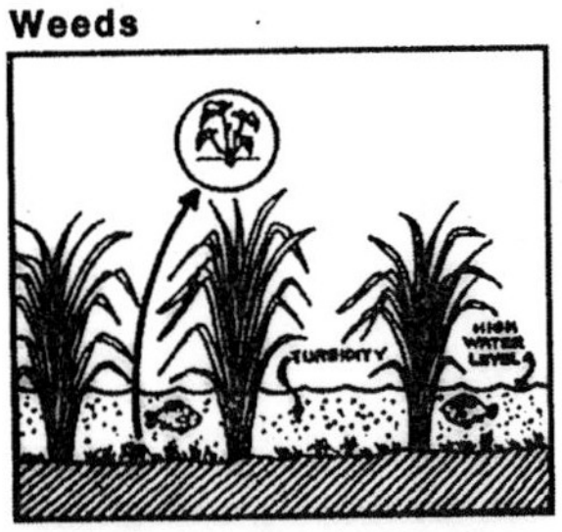

Weeds are controlled by direct feeding, increased water turbidity and constant flooding

Indirect

Leaffolder

Disturbed moths fly up and are preyed upon by birds or other predators

Issues for further consideration

By feeding on rice pests, fish may reduce damage to the rice crop. At the same time, the stocking of fish helps in reinforcing IPM practices that have been the basis for increases in rice yields in many parts of Asia. Farming rice and fish and consciously avoiding the use of pesticides may be an important first step for farmers towards getting better prices particularly at markets near big cities where organically farmed products are of increasing demand.

FISH FEEDING AND MANAGEMENT

Using animal wastes in fishponds

Ruben Sevilleja, Jose Torres,
John Sollows and David Little

Animal wastes flow

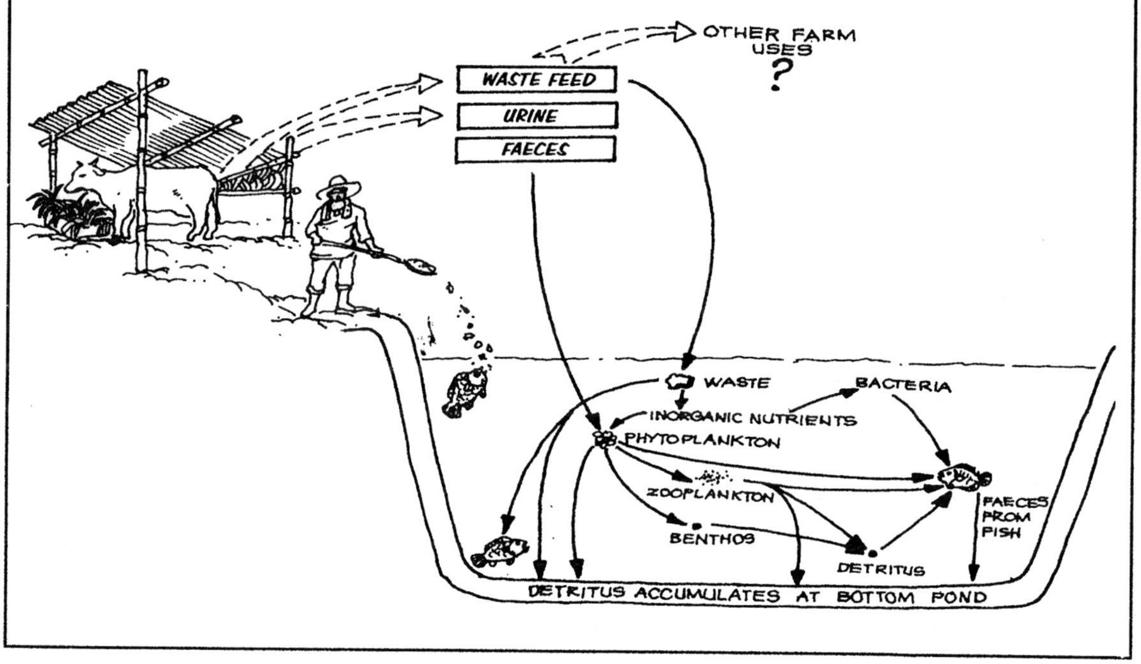

How animal wastes work in a pond

Direct feeding value of pure wastes is known to be poor. Wastes act by:

- stimulating phytoplankton production; and
- acting as substrate for bacterial production (detritus) and as feed for zooplankton.

These two processes are strongly interlinked, since phytoplankton is a major source of detritus for bacterial production. Also, phytoplankton, through photosynthesis, is the chief producer of dissolved oxygen in the pond used by all organisms including fish.

Factors to consider before using animal wastes

1. Are wastes available on-farm? If so, are the wastes already used? Should they be diverted for use in fish culture?

   Livestock wastes are often important as crop fertilizers and fuel. Consider the opportunity costs.
2. Is it worth raising livestock, especially to generate wastes for aquaculture? Consider:
   - costs/difficulties of doing so (e.g. feed availability and cost, marketing difficulties, technical abilities and interest of farmers); and
   - inorganics are now cheaper to use than livestock manure in many places

Management factors to consider

1. Are all wastes to be used in fish culture?

   If wastes are to be used elsewhere, they should be collectible prior to entering the pond (e.g. use a sump). Also, wastes should be available in larger quantities at certain periods when their use should be reduced for fish culture (e.g. during the cool season).

2. Can all wastes be collected?

Feedlot livestock are kept confined at all times so all the wastes can be collected and used.

Small-scale farmers often allow livestock to graze or scavenge during the day and only confine these at night. This reduces feed costs considerably, often allowing only on-farm or low-cost, supplementary feeds to be given. However, collectible wastes will be less.

3. Livestock may be penned at the farmer's house for security or traditional reason; this may limit potential advantages of integration. Labour is required to collect or prepare livestock feed.

4. Ponds may be multifunctional. Large animals are usually denied access to the pond because entry to and wallowing in it can destroy the dikes and cause turbidity which reduces natural food production.

- Livestock wastes vary in terms of both quantity and quality which are affected by the following:
  - food quality of livestock
  - species (monogastrics and ruminants) and size
  - stage in life cycle (breeder, grower, etc.)
  - solids only or mixed with urine
  - amount of waste feed
  - contamination with bedding materials, rainwater, soil, etc.
  - method and period of storage

Layout/design options

On the pond dike

Pens close to the pond to reduce labour cost of loading waste

Over the pond

Pens are cooler and more humid

In the layout/design aspect, consider:
- size and number of livestock;
- space availability/land cost; and
- relative cost of materials

Design the pond to allow limited access.

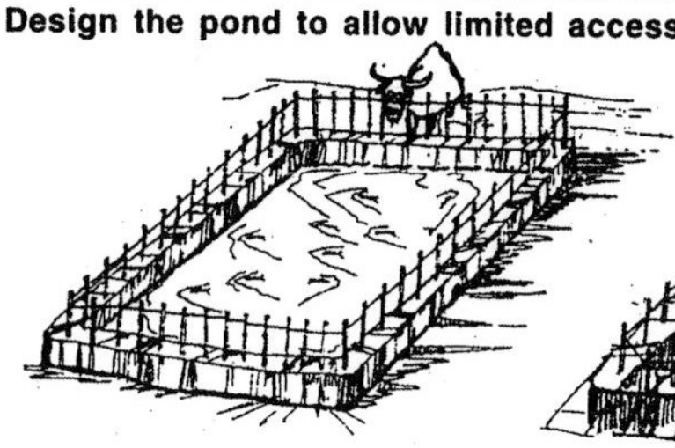

Fence around pond keeps buffalo out.

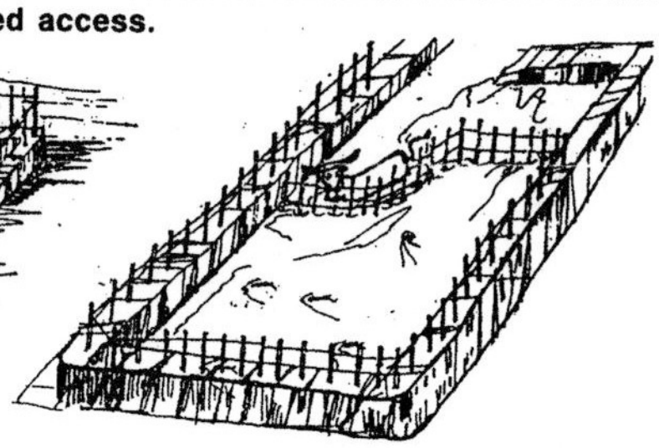

Fence across pond lets buffalo in water.

Pigs and chickens are monogastrics. They are fed a high-quality diet and their waste is high in nutrients.

Buffalos and cows are ruminants. They are given a diet low in nutrients and their waste is low in nutrients. However, they are cheap to feed.

Miscellaneous facts

- Young livestock tend to feed on diets higher in protein so their waste has more nitrogen and is better as a pond input.
- Ruminants' faeces contain high levels of carbon relative to nitrogen and discolour the water. Generally used alone, they give low-fish yields. Consider use of ruminants' urine as it contains a better balance of nutrients.
- Laying hens are fed different diets than broiler chickens and their waste is particularly high in phosphorus.

Tips for proper waste application

- First application can be done about 1-2 weeks before fish stocking to produce natural food for immediate fish consumption.
- Apply or load manure after sunrise (about mid-morning).
- Maintain a regular schedule or routine of application.
- Make sure that freshwater is available for flushing in case of direct oxygen depletion.
- Scrape off 2-5 cm of the pond bottom soil during pond preparation. This can serve as an excellent fertilizer for vegetables.

Water quality management

Too much manure when loaded in fishponds can cause dissolved oxygen depletion resulting in fish mortalities. When manure loading is excessively high, too much decomposition occurs; thus, the biological oxygen demand is high, using up the available dissolved oxygen.

Phytoplankton produces dissolved oxygen during the day but consumes it at night. Another source of dissolved oxygen in a static water is diffusion of atmospheric oxygen.

Indicators of low dissolved oxygen

1. When plenty of fish are on the water surface "gasping for air" (i.e. they are consuming oxygen from the thin and oxygenated top layer of water)
2. When air or gas bubbles are observed in the water
3. When the pond water is brownish or greyish
4. When the pond water smells pungent.

What to do when dissolved oxygen is low

- Stop loading manure.
- Add freshwater into the pond while draining water off the pond bottom.
- Stir the pond water by striking the water surface with tree branches or other appropriate materials; row repeatedly across the pond.
- Make provisions for flow-through system (if water is readily available).
- Use mechanical aerators (if available).

If the water is turbid because of suspended sedimentary particles, spread over the pond surface chopped rice straw or hay, allowing them to settle at the pond bottom together with the silt. But caution: too much decomposing hay can also deplete dissolved oxygen.

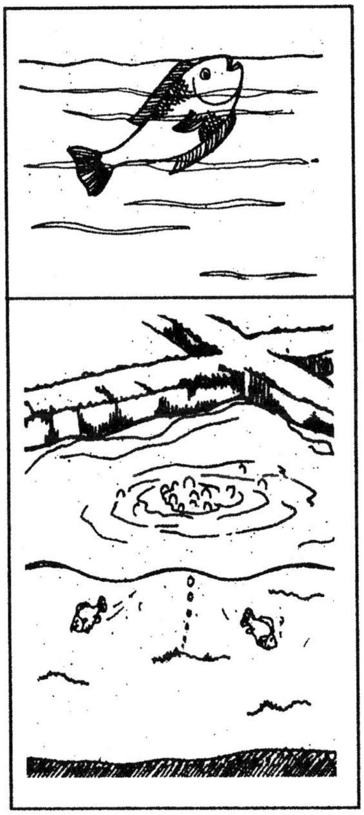

The pH or hydrogen ion concentration determines whether the water is acidic or alkaline. Highly acidic water (4 or below) can result in fish kills.

Methods to measure pH

Use equipment such as litmus paper, pH meter and water quality measuring kit.

Here is a practical method. Test the water: if water tastes sour, it is acidic. Know the water source; acidic water comes from swamps, bogs or stagnant areas.

What to do if water is acidic

- Stop loading manure.
- Apply lime.

Ways of knowing the presence of hydrogen sulphide

Hydrogen sulphide is a poisonous gas emitted from the pond bottom as a result of decaying and decomposing organic matter. Its presence is indicated by the following:

Causes of and possible remedies for different water quality problems

Observations	Causes	Possible remedies
Green water; little surface scum; active fish behaviour	No problem	
Cloudy water; fish appear hungry	Stocking density of fish too high	Harvest some fish; add more fertilizer
Deep ponds; high dikes surrounded by trees	No wind-caused turbulence; poor water circulation	Agitate during critical periods; remove wind breakers; keep water level high
Very green; surface scum; gas bubbles	Overabundance of nutrients for stocking density	Agitate during critical periods; add small amounts of quicklime to precipitate scum; reduce overall nutrient inputs; maintain fish at higher stocking density and/or release plankton-eating fish
Cloudy, still weather	Low light intensity; no natural agitation of water surface	Agitate; add freshwater; harvest some of the fish
Brown, discoloured water; gas bubbles; pungent odor	Overuse of manure	Use less manure and more inorganic fertilizer

- Emission of unpleasant odor resembling that of a rotten hard-boiled egg
- Presence of dead fish also in the source canal

What to do when hydrogen sulphide occurs

- Agitate the pond water.
- Add freshwater.
- Regulate or stop manure loading.
- In serious cases, drain pond and dry pond bottom for 1-2 weeks.

Fish harvesting methods to remove off-flavour

Off-flavour or muddy taste of fish harvested in highly manure-loaded pond and in pellet-fed ponds can be a serious problem if fish farmers do not follow the proper harvesting procedures. People will not buy or eat the fish with off-flavor or muddy taste.

Here are some suggestions to remove the off-flavor or muddy taste:

1. Stop manure loading or delivery to the fishpond at least two days before harvesting.
2. Partially drain the pond leaving about 40-50 cm water depth.

3. Harvest fish by seining before draining the pond totally. This will minimize fish mortality and the murky odor of fish associated with muddy water.
4. Transfer fish to a net enclosure installed in a pond with clean water or in holding tanks with running water and hold the fish for at least 4-6 hours but preferably for several days.
5. Sell fish alive or fresh.

Ways to measure water transparency (or turbidity)

Use of a Secchi disk
The disk is lowered into the water from a calibrated rope. If it disappears within a depth < 30 cm, the water is turbid.

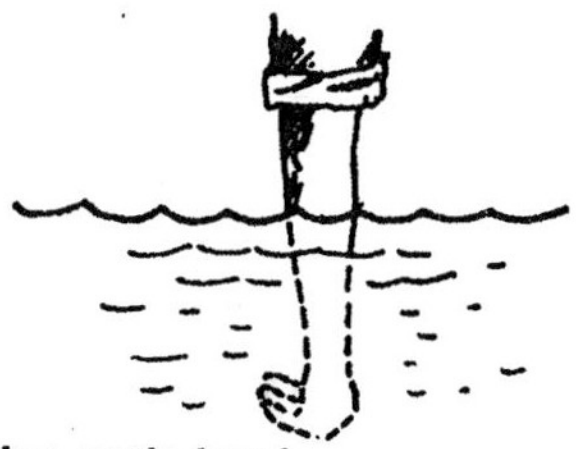

Using one's hand
With the hand stretched forward, cup the palm and bend it towards you. In this position, slowly dip the hand into the water until the palm becomes invisible. Transparency is expressed as the distance from the wrist to the end of the water mark on the arm.

Issues for further consideration

With high loading rates, the pond sediment can be raked periodically during the early afternoon period, when the dissolved oxygen concentration is at its peak, in order to aerate the bottom sediment and facilitate aerobic decomposition processes.

It is wise to apply the desired amount of manure in small doses and more frequently. Daily application gives much better results than weekly and fortnightly applications. It helps in keeping the water quality parameters under control. Frequent but smaller applications do not allow the deterioration of water quality abruptly. As soon as any symptom of algal blooming or dissolved oxygen depletion is visible, the application of manure should be suspended. Application of manure in liquid form (or sewage) keeps the particulate detritus materials suspended in the water column for longer period of time allowing for bacterial decomposition at a higher rate due to the aerobic environment. During this period, these detritus particles are enveloped with a multiplying bacterial population and are also available for fish as well as zooplankton as quality feed material.

Sewage-fed fish culture

S.D. Tripathi and B.K. Sharma

Sewage is a rich nutrient resource, cheaply available around big towns and cities. It can be well-utilized: for fertilizing paddies, fishponds and horticulture crops. Waste recycling also helps in maintaining a clean environment. This paper is based on existing practices in eastern India.

Rice-fish/prawn

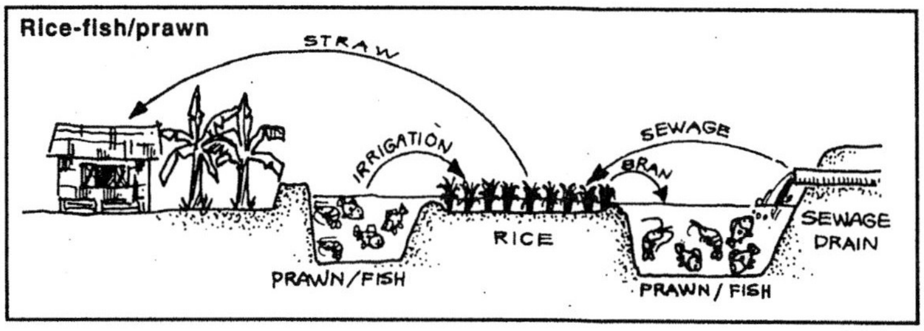

Rice-fish/prawn

In areas where irrigation facilities are not available, a second crop of rice is possible by constructing water storage areas within the field. These could be in the form of lateral, central or marginal trenches or unilateral/bilateral ponds which are also utilized for aquaculture. Based on the input requirements for a 0.4 ha field, the following methods are used by farmers:

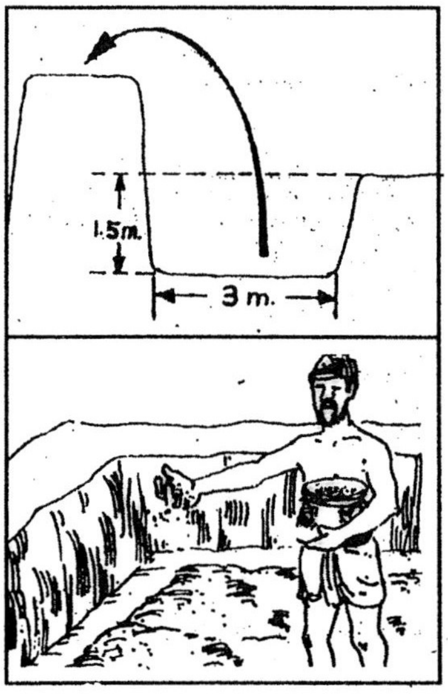

1. Raise the peripheral dikes by digging a perimeter trench (3 m wide x 1.5 m deep) or a lateral pond. If necessary, inlets and outlets are provided and guarded with meshed screens.
2. Fill the trench with sewage water to a level of 15-20 cm.
3. Deepwater rice (e.g. CN 570, 652; NC 487 or 492) is sown directly after the first monsoon shower.
4. When the water level in the trench is about 60-70 cm, stock about 400 mature (1.5-2 g) *mola* (*Amblypharyngodon mola),* a small indigenous species high in vitamin A) together with 8 000 bata (*Labeo bata*) having an average weight of 2 g. As soon as 3-4 g prawn (*Macrobrachium rosenbergii)* are available, 2 000 juveniles are also stocked. The fish and prawn move about the field when the water level in the trench rises and covers the paddy.
5. The water level in the field and the trench falls with the end of the monsoon. The paddy ripens in November/December and about 500-600 kg of deepwater rice are harvested from the field after 150 days of growing. The fish and prawn continue to grow in the trench. Utilize the water in the trench for raising a second crop of rice. Fertilize it by taking in sewage to a level of about 10 cm each month from December to February. A low-level dike is constructed all around to maintain a 10-15 cm water level in the paddy field.
6. The field is fertilized with sewage and seedlings of high-yielding rice varieties (e.g. Ratna or IET 4094) transplanted in January.

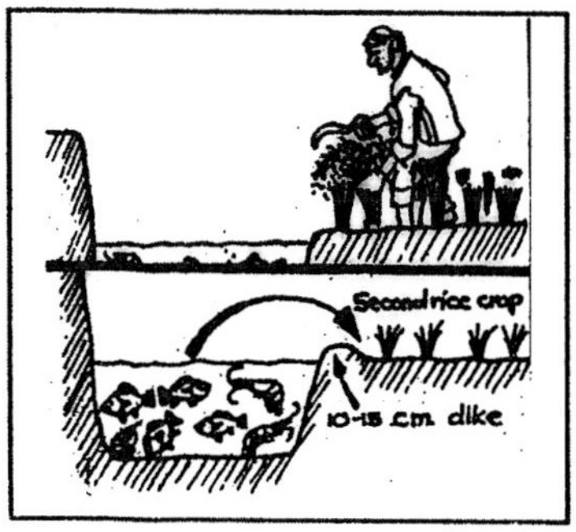

Budget (in rupee) for rice-fish-prawn culture in a 0.4 ha unit

Costs	Rs
For first *(kharif)* crop	
Rice seed (44 kg at Rs3.50/kg)	154
Labour (20 person-days for ploughing, sowing, harvesting and thrashing at Rs18/day)	360
For second *(boro)* crop	
Rice seed (32 kg at Rs3.50/kg)	112
Labour (44 person-days for cleaning, transplantation, harvesting, etc.)	792
Pesticides	80
Fish seed and transport	2 500
Total costs	3 998
Income	
Sale of first paddy crop (560 kg at Rs2.50/kg)	1 400
Sale of second paddy crop (2 240 kg at Rs2.50/kg)	5 600
Sale of 210 kg fish/prawn	6 240
Total income	13 280
Balance	9 282

1996: US$1 = Rs25.50

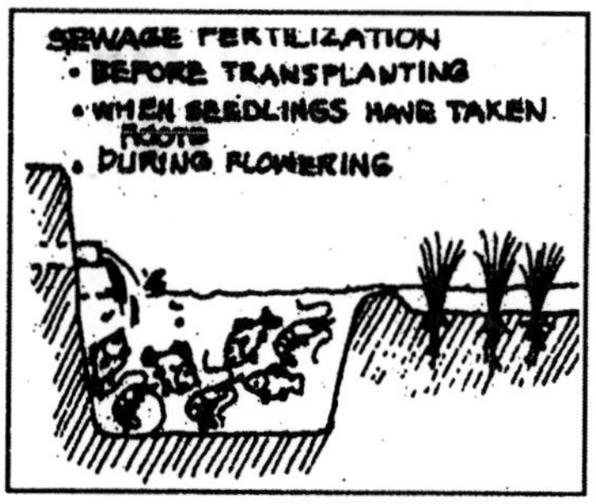

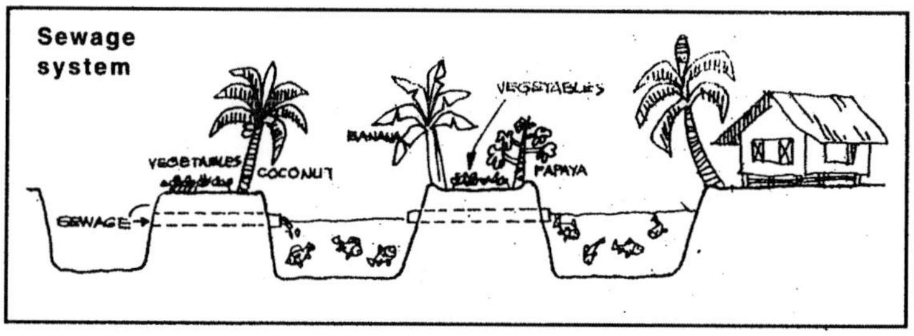

7. Sewage fertilization is repeated when the seedlings have taken roots and again during the flowering stage. The fields are irrigated regularly and the water level is maintained until the rice is mature. Pesticides are used only when necessary.
8. A partial harvest of prawn (50 g), bata (20 g) and mola (20 g) is made.
9. The paddy is harvested in April with a yield of about 2.0-2.4 t.
10. The fish are finally harvested in end April or early May. The total fish harvest is about 112 kg bata, 50 kg prawn and 45-50 kg mola.

Advantages

1. The second rice crop contributes to additional food production, employment and income generation.
2. Fish crop provides a rich protein food of high market value and adds considerably to the farmer's income.

Limitations

1. Trench/pond construction is useful only in water-retentive soils.
2. Difficulties are encountered in fish seed transport, if away from the main road.

Horticulture-fish

The use of sewage for aquaculture and horticulture results in high yields and economizes on fertilizer and feed costs, resulting in higher profits. Based on the input requirements for a 0.4 ha pond, the following procedure is recommended:

1. Broadcast about 200 kg of quicklime over the entire pond surface after it is drained and dried for about 10-15 days.
2. Load the pond with a 30 cm layer of sewage in early June which gets diluted with rainwater and filled up to a level of 1.2-1.3 m in early July.
3. Stock with 3 000 fingerlings of six species (catla, 15; silver carp, 25; rohu, 25; grass carp, 5; mrigal, 20; and common carp, 10) or 2 000 fingerlings of three species (catla, 40; rohu, 30; and mrigal, 30).
4. Use the dikes (500-1 000 m² of land around the pond bank) for growing vegetables, beginning with monsoon crops, followed by

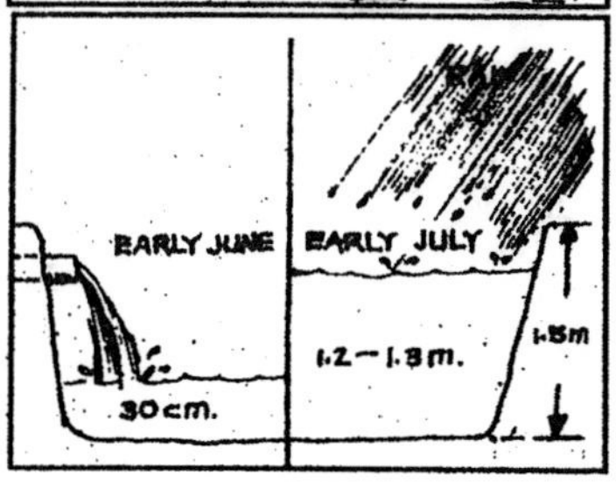

winter and then summer crops. Each crop is harvested as soon as it is ready. About 1 500 kg of vegetables are harvested from 500 m² of dikes. A wide range of vegetables can be planted in simple mixed or multiple cropping: okra, eggplant, cucurbit, gourds, cabbage, cauliflowers, potato, radish, tomato, onion and leafy vegetables like *Amaranthus*, *Ipomoea*, fenugreek, spinach, etc.

5. Load the pond with sewage effluents once a month to the extent of one-fourth or one-fifth of the water level.

   Feed all waste leaves to the grass carp in the pond; 80 kg of leaves give about 1 kg of fish.

6. The pond is netted every 15 days and marketable fish is harvested. A total of 2 400 kg of fish can be harvested from the pond.

Advantages

1. Waste utilization/recycling of domestic sewage brings about a reduction in biochemical oxygen demand/bacterial load before releasing in streams.
2. High-stocking densities and high-yield rates, especially of plankton feeders as well as detritus feeders, are possible.
3. Low-cost fish/vegetable production.

Disadvantages

1. Copepod parasites due to high organic load cause fish mortalities
2. Sudden fall in oxygen level due to cloudy weather or heavy sewage loading also results in mortalities.

Budget (in rupee) for vegetable production on a 1 000 m² plot on the pond banks

Vegetable	Yield (kg)	Production cost	Sales	Net Income
Potato	2 000	1 200	2 500	1 300
Tomato	2 000	1 000	3 000	2 000
Brinjal	2 500	1 000	3 000	2 000
Chili	200 (dry)	1 200	2 800	1 600

Note: About 25 different kinds of vegetables are grown in single/mixed or multiple cropping and an average production of 3 000 kg valued at Rs7 260 is obtained. With the cost of production being Rs5 400, a net profit of Rs1 860 is taken by farmers. In small farms, the farmer uses own labour which accounts for 60 percent of the total production costs; hence, he nets out an income of Rs1 860 + Rs3 240 = Rs5 100 or US$ 204.00 (as of 1992).

Issues for further consideration

It is unclear how widespread the system is in eastern India, or who the main practitioners are. In planning to adopt this system, it is important to study the agroclimatic conditions, and know what the main limitations and constraints are. A widespread system is the polyculture of small Indian major carp and tilapias in the extensive sewage fisheries of Calcutta where there have been major conflicts between absentee landowners who control the sewage-fed fishponds and landless people wanting to plant rice. Most of these fishponds, which used to extend over 40 km², are managed to reduce the risk from heavy metal contamination of sewage and to produce small, live fish for poorer urban markets. This is a parallel activity to the use of solid waste and sewage water for the production of vegetables. This is not carried out by fish farmers and to some extent the horticulturists and aquaculturists are competitive for sewage.

A different notable system is in the Tranh Tri district of Viet Nam where Hanoi's sewage is reused in fish and rice-fish systems. The system was studied by AIT, Bangkok, Thailand, and has further developed and expanded in recent years with some particularly positive impacts on women's access and control of resources.

The system is best suited for peri-urban situations. Experiences suggest that it can be applied for smallholders, medium-sized entrepreneurs and larger commercial farms.

Where necessary and available, lime can be applied in this system to improve and ensure pond productivity.

On health issues in relation to wastewater use, the reader may want to consult WHO (1989) Health guidelines for the use of wastewater in agriculture and aquaculture. Report of a WHO Scientific Group. World Health Organization, Geneva. WHO Tech. Rep. Ser. No. 778, 74 p.

Biogas slurry in fish culture

S.D. Tripathi and B. Karma

Cow dung is commonly used as a fertilizer for fishponds in India but fish production is limited to 1 500-2 000 kg/ha. These yields can, however, be more than doubled if the dung is first fed to a biogas plant and the digested slurry then used instead of the raw dung. The following methodology for a 0.4 ha pond exemplifies the technology.

1. Prepare the pond using the urea-bleaching powder method or by draining-drying in June.
2. Stock the pond with 2 000 (5-8 g) fingerlings of six Asiatic carps: catla, 20; rohu, 25; mrigal, 20; silver carp, 20; grass carp, 5; and common carp, 10.
3. Fertilize the pond daily with 30 litres of biogas slurry. The slurry is rich in nitrogen and phosphorus, and is free from toxic gases which are produced when cow dung decomposes in ponds.

   Excess slurry is used for the field while the gas is used both in the kitchen and for lighting the house. The slurry is not applied on a cloudy day or when the fish come to the surface to gulp air.
4. Surface feeders will have reached about 1 kg individual body weight in 6 months. All marketable fish are then harvested every 2 months and replenished with an equal number of fingerlings. A total of 2 000 kg of fish is obtained using biogas slurry as against 800 kg if raw cow dung is used.

Budget (in rupee) for biogas slurry-based fish culture on a 0.4 ha pond

Operational expenses	
Pond preparation	800
Seed	400
Transport	100
Biogas plant (two units, depreciation value on 5-year life span)	2 000
Lime (80 kg)	400
Netting charges	500
Total receipts	4 200
Sale of 2 000 kg fish at Rs15/kg	30 000
Net income	25 800

1992: US$1 = Rs25.50

Advantages

- Savings on inorganic fertilizers and feed (60 percent of operation costs)
- Environment-friendly, no oxygen demand
- Savings on fuel and electricity
- Cooking with biogas removes drudgery of women for fuelwood collection and helps in keeping the kitchen and environment clean.

Limitation

- Slurry/gas production is poor during cloudy days or when temperatures are low.

Issues for further consideration

Although widely promoted as an appropriate technique for use in aquaculture systems, there is very little evidence that biogas slurries have been used by individual households on a sustainable basis. The systems that were demonstrated were experimental. Biogas digester designs have evolved. The conventional types promoted in India were too capital intensive for adoption by poorer people. The low rates of adoption suggest that these digesters did not meet peoples' needs. Research at AIT in Bangkok, Thailand, suggests that the technical benefits from digesting waste were marginal or negative.

The system, as described may be more useful for commercial-scale livestock farm operators intending to integrate aquaculture into their farming system.

Experiences with biogas digester technology dissemination have been that this is a clustered technology which is best operated with several households sharing one system, to provide adequate levels and continuity of waste input. Recent developments in digester design have made simpler, very low-cost systems available. In southern Viet Nam, where these cheaper types of digesters were promoted and their adoption was monitored, the retention of the technology within aquaculture has been poor. Farmers prefer to use their livestock waste fresh.

Additional benefits of the system will accrue to women by providing biogas for cooking.

Plant sources of feed for fish

S.D. Tripathi and B.K. Sharma

The case study presents two different concepts: (1) the cultivation of aquatic macrophytes concurrently *(trapa)* or in rotation *(euryale)* with fish and (2) the cultivation of terrestrial grass (napier) on the banks of a pond and its feeding to fish.

Aquatic plants

In India, trapa *(Trapa bispinosa)* and *makhana (Euryale ferox)* are two seasonal, aquatic cash crops which are grown extensively in Madhya Pradesh and Bihar, respectively. While the environment is not congenial for Indian carps, common carp goes well with *trapa* and air-breathing fish with *makhana*. Based on the input requirements for a 0.4 ha pond, the procedures to be adopted are given below.

Concurrent *trapa*-fish culture

1. Transplant *trapa (Trapa bispinosa)* seedlings in May/June in a perennial pond. These plants make use of the available organic matter for their growth.
2. Stock 800 (50 g) common carp fingerlings in September-October.
3. *Trapa* fruits ripen in winter and are harvested from November to January. A production of 3-4 t of fruits is obtained.
4. Fish are harvested in April/May when 750-1 000 g fish are available. A total of 400-500 kg fish are harvested.

Trapa bispinosa

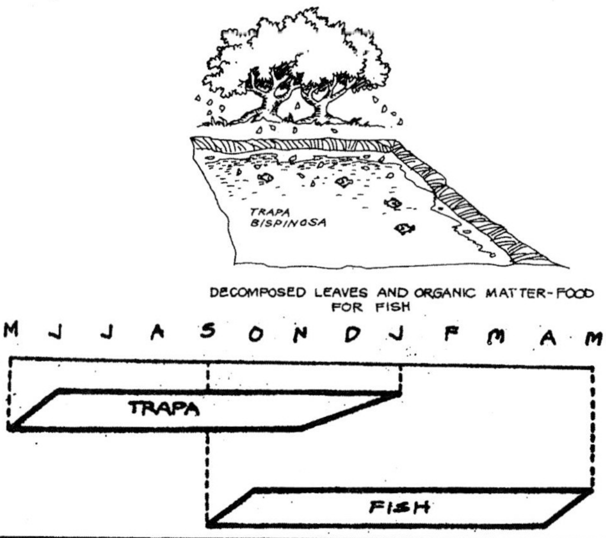

Euryale ferox

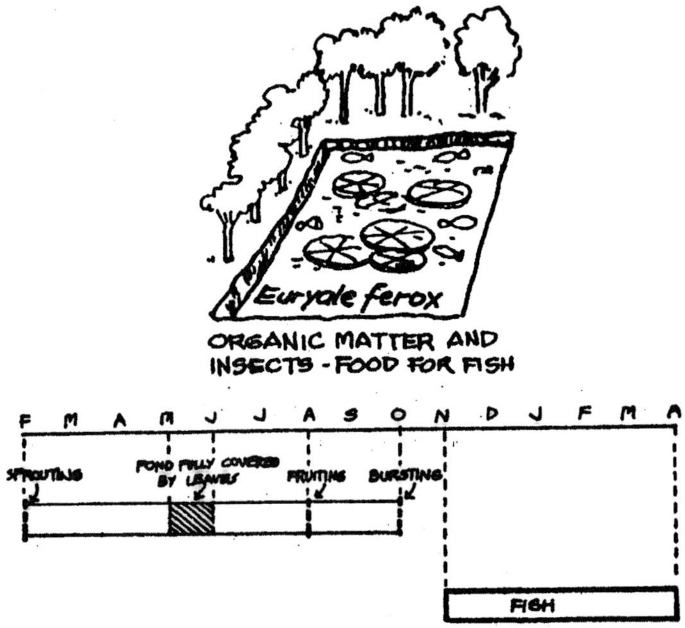

Rotational *euryale*-fish culture

1. The seeds of *Euryale ferox* sprout in February and the leaves cover the pond fully in May/June.
2. The plants start fruiting in August and burst in October, scattering the seeds at the pond bottom which are collected by scanning the bottom mud.
3. Stock 1 200 (8-10 g) air-breathing fish *(Clarias batrachus)* in November and harvest in April; when about 500 kg of fish can be obtained.

Napier grass

Besides aquatic vegetation such as *Hydrilla, Ottelia, Potamogeton,* etc., green grass has an important role in feeding grass carp. Hybrid napier, once sown on pond banks, can be cropped continuously for five years, needing little irrigation during summer. A new system utilizing aquatic vegetation/green grass alone for fish production gives high yields at very low costs. It is labour-intensive and highly suitable for small, shallow ponds (0.06-0.15 ha). Based on input requirements for a 0.1 ha pond, the methods to be followed are given below:

1. Prepare the pond in May/June by using urea-bleaching powder method or by draining, if a source of water for filling the pond is available.
2. Seven to 10 days later, stock the pond with 200 (50-60 g) grass carp. Feed them to satiation with Hydrilla. (System of feeding ad libitum: fish are satiated when they have stopped feeding and there are still some feed material left lying about.) Within about a week, the pond is also stocked with 40 each of catla, rohu, mrigal, silver carp and common carp (5-8 g). Grass carp is gradually weaned from Hydrilla to napier grass.
3. Feeding is done regularly to satiation.
4. Silver carp, catla and common carp will be the first to attain a weight of 1 kg each. From the fifth or sixth month onward, these are harvested one after another. Replenish the harvested fish with an equal number of fingerlings.
5. Hybrid napier is planted at 1 root slip/m^2 and manured with 2.5 t farmyard manure/1 000 m^2. Irrigation is done at 10-15-day intervals. The grass is cut after 75 days, followed by 45-day intervals. About ten cuts can be taken from each plant. A production of 12-15 t napier from 1 000 m^2 is taken. About 2 000 m^2 land area will produce enough napier to feed the fish in a 0.1 ha pond. This means that to provide sufficient grass to feed the fish, twice the pond area is needed for growing napier.
6. About 400 kg of fish can be harvested from the pond in the course of one full year.

Advantages

- Utilization of napier/weeds for fish production at no cost
- Utilization of pond resource for fish production in *trapa/makhana* ponds
- Additional income and employment generation

Limitations

- Nonavailability of large-size grass carp for stocking and lack of appropriate transport for these
- As large quantities of grass are required, napier/weed integration is possible in small ponds only.

Issues for further consideration

In Asia, the concurrent production of aquatic macrophytes with fish is fairly widely used. Farmer motivation for using their ponds for this purpose is not well understood. Sometimes seasonal constraints prevent the continuation of year-round fish culture, or the plant has a seasonal production period or market. In Binh Chan, near Ho Chi Minh City in southern Viet Nam, lotus is commonly rotated with breeding tilapias in shallow ponds fertilized with sewage during the off-season for tilapia seed production.

Knowledge about the prevalence of the napier grass system today would be valuable as this is a resource-intensive system. Where grass carp are raised as a central part of the resource-poor farmer systems, they normally have to source a high proportion of the feed off-farm from common property areas (e.g in northern Viet Nam). This raises questions of competition between livestock owners and fish farmers, individual households and the community as a whole.

If the focus is on the production of large individual fish, these systems would be used by resource rich farmers who can risk such systems, probably using employees for the considerable labour requirements. There are considerable implications for adoption of vegetation-based systems by poor farmers especially with regard to their labour and how this is organized within the household, even though it appears to be simple.

Depending on local species, different plants may have to be tried for use as direct fish food or for making green water (i.e. pond fertilization).

Fish Breeding and Nursing

Carp breeding using off-season wheat fields

S.D. Tripathi and B.K. Sharma

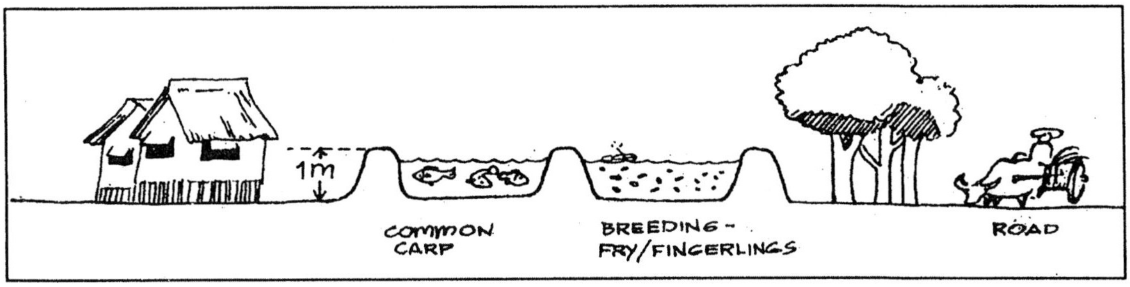

About 300 000 ha of wheat fields around Jabalpur, Madhya Pradesh in India are virtual rainfed ponds (*havelis*) from July to October. There being no source of irrigation, rainwater is impounded in these fields (with about 1 m high dikes) until the onset of winter when they are drained, ploughed and the wheat is sown. This period lasts for 3-4 months wherein the field is used for common carp seed production. Based on the input requirements for a 0.4 ha field, these procedures could be followed.

Seed production

1. Select a field near the road but away from a floodprone zone. Check the dikes and put meshed screens on the inlets and outlets, if provided.
2. Spray an emulsion of 20 litres diesel and 7 kg of cheap washing soap on the water surface to kill predatory aquatic insects as soon as about 60-80 cm of water gets accumulated in the field.
3. After spraying, release four healthy and fully ripe females along with an equal number of males, each weighing about 1 kg. Provide 2-3 kg of Hydrilla or Eichornia at three or four places in the field. A fully ripe, healthy female can be distinguished by its swollen, bulging abdomen and

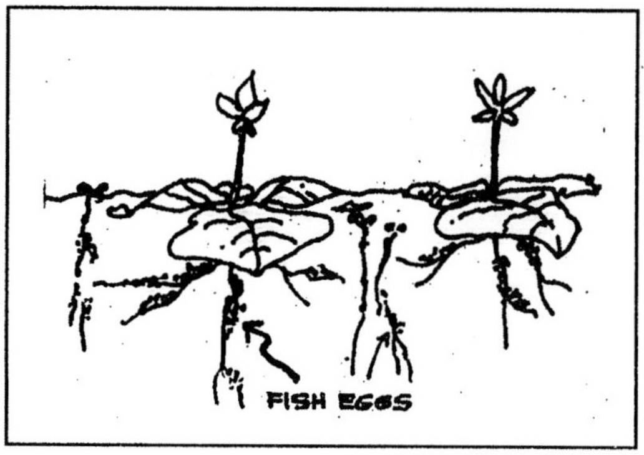

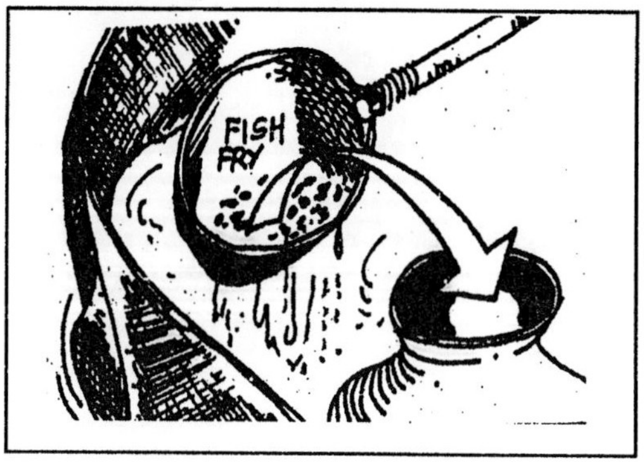

a reddish genital region which is pit-like in the male. The males also ooze milk with gentle pressure on their abdomen.

4. The fish breed within 24-48 hours of stocking or take a day or two more, if they are not fully ripe. The eggs are laid on the weeds and hatch within 48-72 hours.
5. Harvesting can be done after 15-20 days. Approximate yield is 100 000 fry of about 25-30 mm size. If the field is fertilized with 2 000 kg cow dung and the fish are fed with an artificial feed comprising groundnut oil cake and rice bran (1:1 by weight), the survival is high and the growth is fast.
6. The remaining fry attain a size of 40-60 mm by the time the fields are to be drained when these (about 20 000 fingerlings) are also harvested.

Budget (in rupee) for carp in 0.4 ha *haveli* wheat fields

Costs	
Cost of 8 kg live broodfish (4 females and 4 males, 1 kg each) at Rs25/kg	200
Transport	200
Soap oil treatment	125
Total costs	525
Income	
Sale of 50 000 fry at Rs10/1 000	500
Sale of 20 000 fingerlings at Rs100/1 000	2 000
Sale of 6 kg of fish at Rs15/kg	90
Total income	2 590
Balance	2 065
Net income/ha	5 162

Issues for further consideration

In considering the adoption of this technology, further information may be required. Who buys the seed and in which quantities? What are the alternative sources of fish seed (if any)? What role do the different members of the household have in the activity? Where do farmers producing the seed obtain their broodfish from? What type, rates and methods of fertilization and feeding are needed? In estimating the economics of the system, cost for these would have to be included.

The technology described here is used in certain parts of India. It can also be adapted to areas with demand for carp seed and can be performed by smallholder farm families to enhance their income.

Nursery system for carp species

Md. Golam Azam Khan

A nursery is a facility where fish seed (spawn, hatchlings or fry) can grow. Efficient fishpond culture requires special preparation of nurseries for receiving spawn and hatchlings. The ideal size of a nursery is 0.02-0.05 ha with a depth of 1.0-1.5 m. The following is an example of a 0.02-ha nursery as prepared in Bangladesh for 15-20 days.

Pond preparation

- Remove all aquatic weeds (day 1).
- Drain and dry the pond.
- Apply 5-6 kg lime/200 m² which helps release food nutrients available and kills pathogenic organisms in the pond (days 3 to 16).

- Refill water if necessary and fertilize (day 19).
- To determine if enough food organisms (plankton) for the fish fry have developed in the pond, the most basic and reliable test involves filtering approximately 50 litres of water through a fine mesh net or cloth into a 2.5 cm diameter specimen tube. Alternatively, a very simple field test in nonmuddy water is to dip one's hand in the water to the elbow. If the hand is no longer visible, the plankton is probably sufficient.

Stocking

- Apply 80-100 g insecticide (e.g. Dipterex) at least 20-24 hours before stocking to kill the back-swimmers or other aquatic insects in the pond (day 29).

- Stock 60 000-70 000 hatchlings/200 m² of 4-5 days age (200-250 g). The hatchlings should be of same age, uniform size, vigorous and released either in morning or late afternoon (day 30). Common carp seed is available throughout the year in Bangladesh.

Fish seed	Availability
Common carp	January - March
Silver carp	February - August
Rohu, mrigal	April - July
Catla	May - July
Grass carp	May - August
Silver barb	March - May

Before the hatchlings/fry are introduced to a new environment, it is important that the temperature inside the plastic bag is approximately the same as the pond water.

Place the bags, unopened, in the pond for 10-15 minutes. Open slowly and introduce small quantities of pond water to equalize the temperature. The fry is now allowed to swim into the pond.

Feeding

- It is often difficult to maintain a high level of natural food for growing fry so that supplementary feeds become necessary (day 31). A mixture of finely powdered oil cake (soya beans, mustard, etc.), rice or wheat bran and fish meal in the ratio of 5:4:1 is to be supplied to fry, at 200 g divided into several portions per day.

Fertilization

- Fertilization is necessary to maintain an adequate level of natural food organisms in the pond.

Care of fry/fingerlings

- Check the pond daily and see if there is an excess of green algae. If so, discontinue application of supplementary feeds. Remove frogs/snakes, if any, in the pond. Increase feed by 10 percent of the rate mentioned above if the growth of the fish is not found steady and good.

Harvesting and transportation

- Harvest the fry/fingerlings (day 60) by using a fry catching net, either in the morning or late afternoon and keep them in the enclosure (hapa) or cistern at least 3-4 hours before transportation (day 60). Transport the fry/fingerlings in oxygenated plastic bags.

  Before transporting, it is important that fingerlings are conditioned. The principle behind this is that they have time to empty their guts before being packed in high densities, so that the pollution of the water in the transport container through excreta is reduced. Clean water from a tube well should be used for conditioning the fingerlings.

Traditionally, young fish are transported in clay or aluminum pots. Recently, the use of plastic bags with compressed oxygen is becoming more widespread, as this allows the fish to be transported in higher densities and longer distances with substantially less mortality. Approximately, 5 litres of water and 15 litres of oxygen are placed in each bag (assuming plastic bag volume is 20 litres).

Species	No./litre
Rohu	50
Bighead	50
Catla	33
Silver carp	60

Budget (in Taka) for a 0.02 ha nursery pond preparation for fingerling production

Costs	
Draining/refilling or poisoning of pond	75
Lime (5 kg)	25
Cattle dung (200 kg)	100
1.75 kg urea and 2 kg triplesuperphosphate	20
Dipterex 0.2 kg	80
60 000 carp hatchlings	600
Supplementary feeds: 20 kg mustard oil cake + 10 kg rice or wheat bran and 4 kg fish meal	180
Netting, labour and others	400
Total	1,480
Income from sale of 30 000 (3.5 – 4.5 cm) fry/fingerlings	3 000
Balance	1 520

1992: US$1 = TK38

Issues for further consideration

From experience, farmers who adopt these techniques are relatively resource-rich and/or have had access to specialized information. Options for sanitation of nursery ponds (e.g. to eradicate aquatic insects that prey on fish larvae) that do not involve the use of dangerous chemicals should be used, e.g. higher levels of quicklime, eradication of pest habitat and continual trapping and removal of pests, etc.

Many nursery operators do not have access to tube-well water, and to condition fish under this situation, farmers may need more detailed information. The conditioning process has more important details, such as what exactly is done and what fish are in good condition and suitable for transporting.

The role of women may be very marginal in Bangladesh nursery systems (from where this example is taken) but in other locations in Asia they represent a major part of the labour force. The methods described are broadly similar to those used elsewhere by commercial nursery operators although they tend to be most intensive in Bangladesh. There they also use the practice of fry and fingerling production in the same pond (one-tier system) in bigger ponds.

The practice of fish seed rearing is followed by a large number of small-scale farmers in Bangladesh, who themselves do not necessarily grow fish to market size. This part-time occupation may be an option for other farm families in areas where the availability of fish seed is still a limitation in the popularization of aquaculture.

Fry nursing in rice-fish systems

David Little, Nick Innes-Taylor,
Dennis Turongruang and John Sollows

Most northeast Thai rice-fish culturist cannot control predators in their fields and finding seed fish over 7 cm in length is difficult, if not impossible Therefore, culturing small fry in a nursery where they can grow, safe from predation, to a size when most of them can escape predation, is often advisable.

Nurseries come in several varieties:

- small nursery pond in or near the field;
- small rice-fishpond, well-supplied with water; and
- nursery cage in larger pond.

Nursery ponds

A small pond, usually 100 m^2 smaller, is most common. During dry season, the pond is dry or cried. Lime and manure are commonly added at about 3 kg and 10 kg/100 m^2, respectively. With the first rain, in new ponds, these rates should often be increased.

Once water begins accumulating, depth and color should be monitored. Is the water turbid? Adding more manure, straw or other fertilizers may help clear the pond. If water is very clear, similarly, fertilizer should be added. This fertilizing in preferably clear water should lead to the growth of plankton, which gives the water a brownish to greenish (preferably) color. Checking the amount of plankton in the water is easily done by observing at what depth the palm of the hand disappears. Ideally, the palm will become invisible around elbow depth. If the palm disappears in 10 cm of the surface, the water is too rich. Fertilization should be reduced or stopped and some new water added if possible.

The depth of water should preferably reach 70-80 cm prior to stocking. The farmer should feel reasonably sure that water should remain near this depth, as well. Two to three thousand 2-cm fry can be stocked in a 100 m^2 pond, for culture up to fingerling size (5 cm). This is enough for a 1 ha field when fish are not fed. If the field is about 1 000 m^2, a 10-20 m^2 pond will be large enough.

Following stocking, the pond can be fertilized and fed with fine, available materials (rice bran, termites, leftovers). Feeding will be especially important in turbid ponds.

Early every morning, the pond should be checked to see if fish are gaping. This is a sign of insufficient oxygen and, if noticed, should be dealt with according to the situation.

Fish are usually held in the pond until the rice is well-established in the field (with 2-3 new tillers) and the fish have reached a length of around 5 cm. This usually takes about six weeks. If there is stable, standing water in the field at this time, fish can be released. If large fingerlings are stocked before rice is well-established, they will damage the rice. (See paper on stocking for rice-fish culture, this volume.)

A nursery allows the farmer to stock fish earlier, thereby prolonging the growing season for the fish and possibly allowing purchase of a wider choice of fish than later in the season. A good nursery also assures higher survival for fry than would be the case in a ricefield, when predation is uncontrollable. The farmer who is used to buying fingerlings will save money by investing in smaller fry.

A bad nursery, however, is worse than no nursery at all. If predators are present, seed fish cannot escape and mortalities will be very high. For similar reasons, pollution due to overfeeding or toxic chemicals can be dangerous. Overheating, particularly in very shallow water, can be another problem. A small patch of shade over the water may be needed. In this case, this should cover only a little of the water surface, since sunlight is needed to produce oxygen and natural

feed. When an existing pond is used to hold water or fish all year round, it should not be used as a nursery. The farmer will do better to dig a small, shallower pond or cage to set a nursing cage of fine mesh in the existing pond. Fish stocked in such a cage will need daily or twice-daily feeding with good quality food.

Nursery cages (hapa nets)

The Aquaculture Outreach project of the Asian Institute of Technology (AIT) has developed

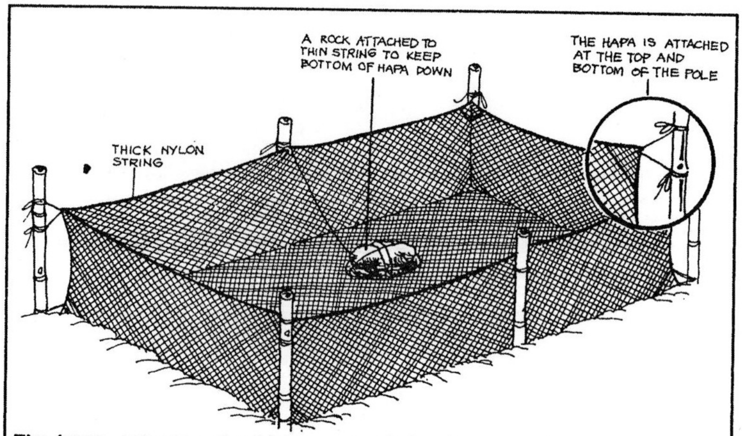

The hapa net cage should be suspended using enough bamboo poles and nylon string. The bottom of the net is kept down using a rock attached to a string for easy removal.

Nylon hapas can be made by hand but are usually stronger when a machine is used. Attention should be given to making the reinforced corner.

bamboo poles for peg

nylon thread

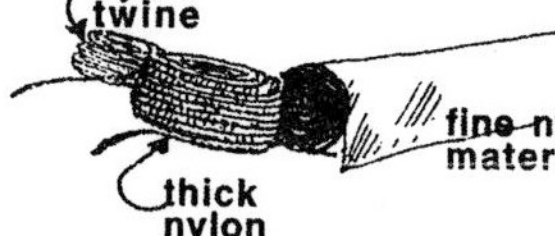

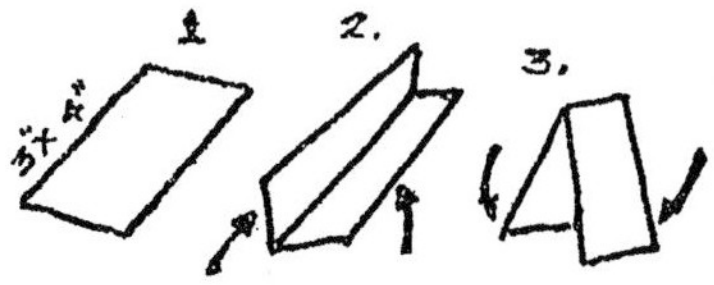

Cut and fold a piece of nylon 3x4 in before sewing and using as a hoop.

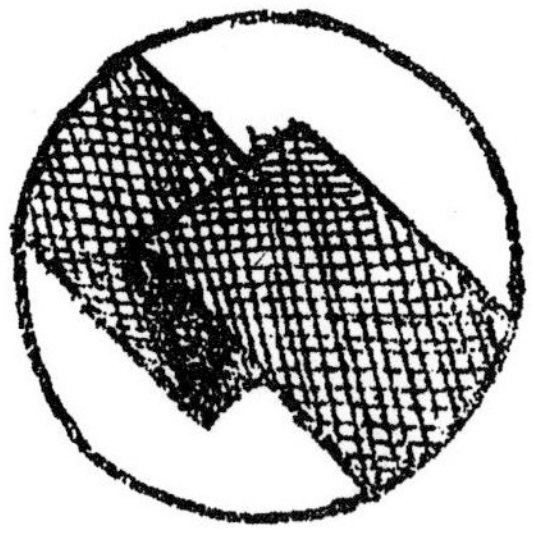

The corners of the hapa should be folded over and double sewn for extra strength. The net material is 90 cm wide.

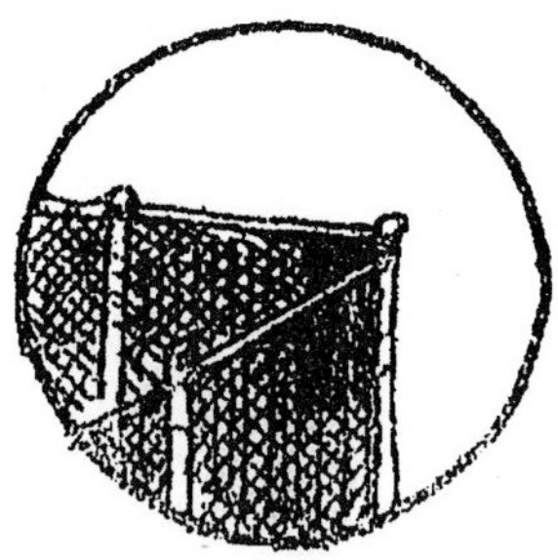

The corners and hoops need to be strengthened by using 3x4 in nylon pieces.

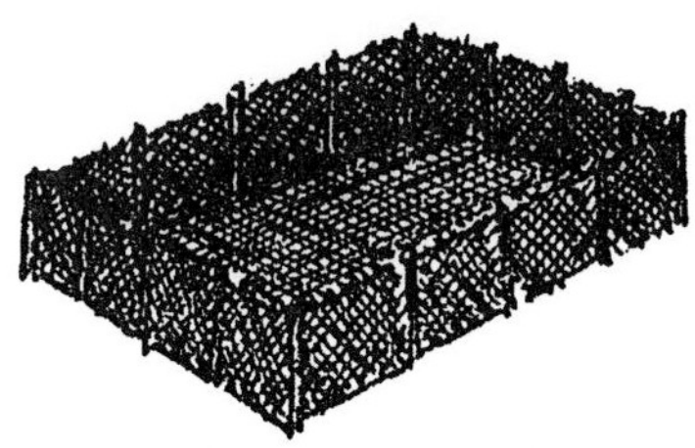

Hoops are required to attach the hapa at the bottom and top of the bamboo pole.

Scrubbing hapa net

After use the hapa should be cleaned and dried to avoid damage by rodents.

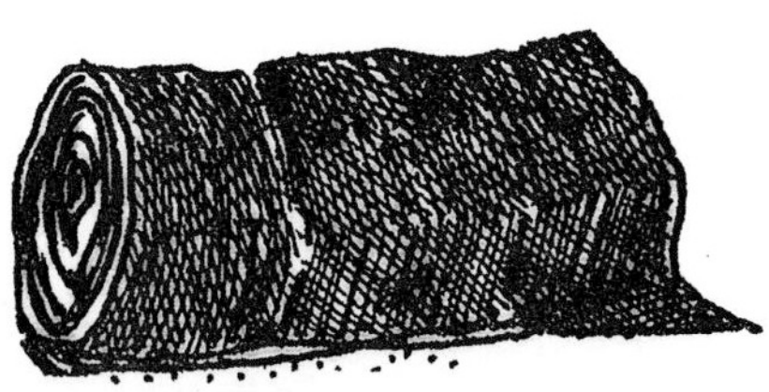

Storage of hapa net

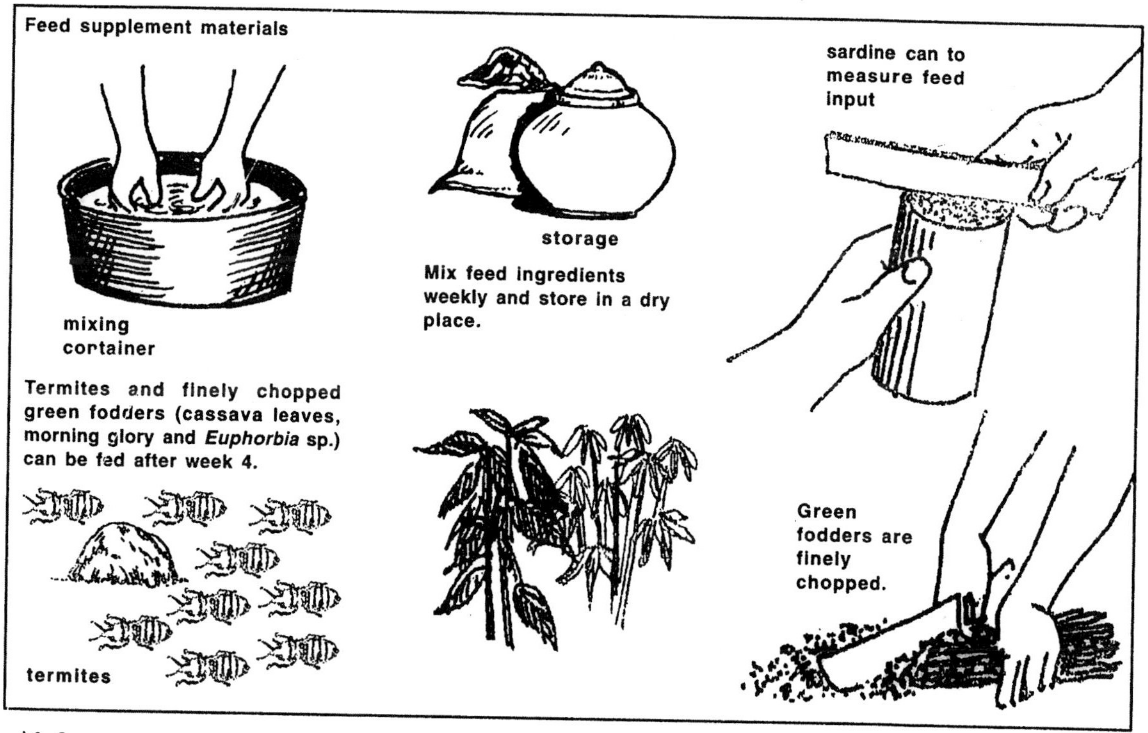

with farmers a nursery cage technology which is becoming popular in northeast Thailand. These fine mesh cages assure an absence of predators, make management easy and give the farmers a chance to become more familiar with their fish. Feeding, however, become, more expensive.

Seed should be given twice daily as a mixed dry mash of duck or pig concentrate (40 percent crude protein) and fine rice bran (at a ratio of 2:1 by weight). This may appear rich, but has been found appropriate in trials with farmers. Feed can be mixed for 1 week and kept in a dry place.

Nile tilapia is best raised in monoculture, while common carp, mrigal, grass carp and silver barb all grow well in monoculture or polyculture; they reach 6-10 cm in 6-8 weeks.

Fertilization of the pond or ricefield using urea and buffalo manure will improve growth and allow some reduction in quantity of concentrate given.

Change feeding rate weekly. Amount of feed should increase and be equivalent to:

10% body weight/day	:	weeks 1 and 2
8% body weight/day	:	weeks 3 and 4
5% body weight/day	:	weeks 5, 6, 7 and 8

Other feeds can be given after week 4.

The feeding unit should be readily available to all farmers. In this case, a sardine can was used.

Issues for further consideration

After the hapa method was developed in northeast Thailand, it has been promoted in areas such as the Lao PDR where there is little access to concentrate feed. Instead, a whole range of supplementary feeds have been used with success.

Hapas are best used in waterbodies that already have fish in them as they help control fouling of the net that can build up on the outside of the cage in the absence of fish. The hapa can also be used for spawning tilapias and small carp, and for holding food fish before marketing them alive.

Local seed production in hatcheries is desirable to both the nursing farmer and his customer, the food fish farmer. Farmers with hapas may also nurse enough fish for their own purposes. Small hapa units ensure that entry costs and risk are low. Hapas have been used successfully to breed and nurse local and improved tilapia strains in northwest Bangladesh and the Lao PDR. In the former, the technology, when promoted for use in backyard ponds, has allowed women to be much more involved than in other forms of aquaculture.

The economics need to be considered, based on the local situation, aside from the risks and constraints.

Fingerling production in irrigated paddy

Francisco Noble

Supply of fish fingerlings is scarce and usually expensive in northern Bangladesh. It is also tedious for fish farmers to procure fish seeds. An alternative is the production of common carp *(Cyprinus carpio)* fingerlings in different types of *boro* or irrigated paddy. Although production figures are below the commercial rates, small farmers can grow their own fingerlings at minimal cost.

The four types of irrigated paddies for fingerling production (as practiced by farmers in northern Bangladesh) have the following features:

Soil :	clay loam
Water supply :	irrigated
Water depth :	maintained at 7.5-10 cm
Area range :	12.5-1 320 m²
Fish species :	common carp
Rice variety :	Bangladesh rice (BR)-3, BR-8, BR-9, BR-14; Tayap and China Pajam

Advantages

1. Fish hatchlings of fry can be reared in different types of boro paddy plot designs without altering the farmer's normal practices in rice production.
2. Farmers' existing resources can be used.
3. Only minimum additional expenses are required.
4. When the fish get bigger than 2.5 cm, they control weeds, pests and insects in the paddy
5. Fish faeces serve as fertilizer for rice.

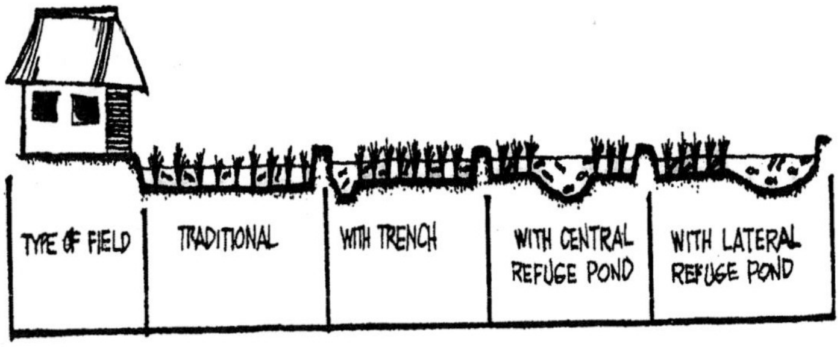

Calendar of activities in fish production

Day 1

Prepare paddy.

7

Transplant rice.

10

Apply approved chemical to kill predators.

11

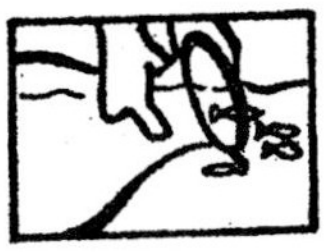

Stock with 2-week old fingerlings at 10/m².

18

Monitor fingerling growth by weighing every week. Maintain water depth at 7-10 cm in the paddy.

53

Harvest the fingerlings. Condition them in hapa before selling, or stock them for further growth.

Note: Apply pesticides for the rice crop only if needed. Remove weeds along paddy embankments to get rid of breeding places of and shelter for, predators.

6. Additional income can be provided.
7. Farmers can sell fingerlings when prices are highest.

What to consider before adapting the technology

1. Paddy soil should have good water-holding capacity.
2. Common carp is recommended for stocking for these reasons:
   - it spawns earlier;
   - fry are available at the same time as *boro* rice transplantation; and
   - it is a hardy fish.

   Tilapia can also be stocked.
3. Fish have higher survival rates in smaller paddies.
4. If possible, use fry (instead of hatchlings) for stocking because they have higher survival rates.
5. Use of supplementary feed, like rice bran or wheat bran, can help increase fingerling production at very minimum cost.
6. To reduce risks of paddy field drying out, use treadle pumps to pump water.

Production details between stock of fry and hatchlings (based on a 1 320 m² paddy)

	Fry	Hatchlings
Stocking rate/m²	10	40
Stocking (no.)	13 200	52 800
Survival rate (%)	40	5
Growth cycle (days)	42	60
Yield (no.)	5 280	2 673

Partial (in Taka) budget for common carp fingerling production in *boro* paddy (based on a 1 320 m² paddy)*

	Fry	Hatchlings
Costs	1 075.64	474.20
Fry cost (TK0.08/u)	1 056	-
Hatchlings (TK00875/unit)	-	462.00
Family labour	-	-
Interest on operating cost (16% per annum)	19.64	12.20
Income return		
Fingerlings (TK0.5/unit)	2 640	1 336.50
Balance	1 564.36	862.30

* Records from ODA/CARE/BRAC Rice/Fish Pilot Project, Rangpur (1991)

Comparative budget and income (in Taka) from three systems: rice, rice and common carp hatchlings, and rice and common carp fry*

	Rice	Rice and hatchlings	Rice and fry
Rice production			
Cost			
Inputs	684	601	601
Labour	492	509	509
Irrigation charges (Tk500/bigha)	500	500	500
Interest on operating costs (16% per annum or 1.33%/day)	89	85	85
Income	3 855	3 943	3 943
Return for rice	2 090	2 248	2 248
Fingerling production			
Cost			
Fingerlings (Tk0.50/pc)	NA	462	1 056
Family labour			
Interest on operating costs (16% per annum or 1.33%/day)		12	20
Income	NA	1 337	2 640
Return for fish	NA	863	1 564
Return for plot	2 090	3 111	3 812
Total	9 800	13 671	16 978

* Records from CARE/ODA/BRAC Rice/Fish Pilot Project, Rangpur (1991)

Notes:
1. US$1 = TK38
2. 1 320 m² = 1 bigha, a standard area measurement used in Bangladesh
3. NA = not applicable

Issues for further consideration

This approach has expanded dramatically in northwest Bangladesh since the case study was written. CARE-Bangladesh have promoted both spawning and nursing of common carp using a farmer field school approach with good success. They promote stocking of fertilized eggs produced by the farmers themselves in the ricefield. Farmers in Rangpur district have sustained their use of the technique to meet their own fingerling needs, although there seems to have been little expansion into any commercialization.

In 1999, the production of tilapia seed in dry season *boro* ricefields was tested using a similar approach to that tested in Thai Binh province, northern Vietnam. There are numerous advantages to this approach for tilapia production, not least of which is a match between timing of fish seed supply with demand early in the monsoon season. In Bangladesh, this method has been beneficial in important fish culture areas in which tilapia production has been constrained by lack of seed.

The described method of breeding common carp has been traditionally practiced in northern Laos and Viet Nam.

Bibliography

Ahmed, A.K.M.M. 1986. An economic analysis of coastal shrimp culture in a mixed farming system, Chittagong-Cox's Bazar Region, Bangladesh. *In* J.L. Maclean, L.B. Dizon & L.V. Hosillos, eds. *The First Asian Fisheries Forum.* p. 153-156. Manila, Philippines, Asian Fisheries Society.

All, A.B. 1990. Rice/fish farming in Malaysia, a resource optimization. *Ambio,* 19(8): 404-408.

Anon. 1986. Integrated farming systems, training for West Bengal fisheries officers. *Bull. CIFRI Barrackpore,* (48): 106 p.

Anon. 1988. Rice-fish culture: Status and potential for increased production in the southwestern states of India. *Indian Fish. Assoc.,* 18: 103-108. 1988.

Baconawa, E.T., Parawan. O.O., Bautista, G.A., Ovalo, H.B., & Catbagan, D.P. 1987. A pilot project on integrated livestock-fish-crop farming in the Southern Philippines. *Resour. Conserv.,* 13(2/4): 265-272.

Banerjee, R.K. 1986. Waste loading, oxygen balance and production efficiency in the integrated fish farming system. *Bull. Cent. Inland Fish. Res. Inst., Barrackpore,* (48): 18.1-5.

Bhaumik, U. 1986. Role of extension in purposeful dissemination of integrated farming system. *Bull. Cent. Inland Fish. Res. . Inst., Barrackpore,* (48): 20.1-6.

Bimbao, M.P., Cruz, A.V. & Smith, I.R. 1990. An economic assessment of rice-fish culture in the Philippines. *In* R. Hirano & I. Hanyu, eds. *The Second Asian Fisheries Forum.* p. 241-244. Manila, Philippines, Asian Fisheries Society.

Bocek, A. 1982. Rice terraces and fish, integrated farming in the Philippines. ICLARM Newsl. 5(3): 24.

Castell, J.D. 1989. An integrated fish farm in China. *World Aquacult.,* 20(3): 20-23.

Chattopadhyay, G.N., Biswas, C.R., Ghosh, A., Chakraborti, P.K. & Bandopadhyay, A.K. 1987. A study on rice-fish culture in coastal saline soils. *J. Indian Soc. Coast. Agric. Res.,* 5(1): 245-249.

Chattopadhyay, G.N., Ghosh, A., Biswas, C.R. & Chakraborty, P.K. 1988. Rice-fish culture in high rainfed coastal saline soils. *In* M. Mohan Joseph, ed. *The First Indian Fisheries Forum.* p. 135-137. Mangalore, Karnataka, India, Asian Fisheries Society, Indian Branch.

Chen, Y. 1989. Animal raising and plant cultivation on an integrated fish farm. *NACA Tech. Man.,* (7): 217-254. Wuxi, China, Network of Aquaculture Centres in Asia.

Clonts, H.A., Jolly, C.M., & Alsagoff, S.A.L. An ecological foodniche concept as a proxy for fish-pond stocking rates in integrated aquacultural farming for Malaysia. 1989. *J. World Aquacult. Soc.,* 20(4): 268-276.

Comia, O.Z. & Mercene, E.C. 1982. Pig cum tilapia farming. *Fish. Newsl. Bur. Fish. Aquat. Resour. (Philipp.),* 11(3): 50-52.

Costa-Pierce, B.A., Atmadja, G.W., Effendi, P. & Zainal, S. 1988. Integrated aquaculture systems in the Saguling Reservoir, West Java, Indonesia. *In* S.S. De Silva, ed. *Reservoir fishery management and development in Asia.* Ottawa, Ontario, Canada, International Development Research Centre.

Das, M.K. 1986. Poultry farming and use of litter in integrated fish farming. *Bull. Cent. Inland Fish. Res. Inst., Barrackpore,* (48): 7.1-10.

Das, N.K. 1986. General constraints in integrated farming systems and their remedies. *Bull. Cent. Inland Fish. Res. Inst., Barrackpore,* (48): 15.1-7.

Datta, S.K., Konar, D., Banerjee, P.K., De, S.K., Mukhopadhyay, P.K. & Pandit P.K. 1985. Prospects for increasing food production in India through different systems of paddy-cum-fish culture in fresh water areas, a case study. Agenda Item No. 7 (IRC/7c3). 12 p. Paper presented at the International Rice Commission Sixteenth Session, Los Baños, Laguna, Philippines, 10-14 June 1985.

Datta, S.N. 1981. Fish/rice coexistence. *Seafood Export J.,* 13(5): 9, 11, 13, 15-16.

Dela Cruz, C.R. 1989. Fingerlings production trials in rice fields in north Sumatra, Indonesia. *In* E.A. Huisman, N. Zonneveld & A.H.M. Bouwmans, eds. *Aquaculture research in Asia: management techniques and nutrition,* p. 97-109. Wageningen, Netherlands, Center for Agricultural Publishing and Documentations Pudoc.

Dela Cruz, C.R. 1990. The pond refuge in rice-fish systems. *Aquabyte,* 3(2): 6-7.

Dela Cruz C.R. & Lopez, E.A. 1980. Rotational farming of rice and fish in paddles. *Fish. Res. J. Philipp.,* 5(1): 39-52.

Din, J., Guo, X., Fang, X., Liu, M. & Zhang, W. 1990. *Effect of animal manure application in fish pond on the bacterial diseases of fish and the food hygiene.* Wuxi, China, Chinese Academy of Fisheries Sciences, Freshwater Fisheries Research Center.

Edirisinghe, U. 1989. Effects of "veluwe" ducks on *Oreochromis niloticus* recruitment under extensive fish culture conditions. *In* E.A. Huisman, N. Zonneveld and A.H.M. Bouwmans, eds. *Aquaculture research in Asia: management techniques and nutrition,* p. 116-120. Wageningen, Netherlands,

Edwards, P. 1982. Integrated fish farming in China. *ICLARM Newsl.,* 5(3): 16-17.

Edwards, P. 1985.. Pigs over fish-ponds. *Pig Int.,* 15(9): 8-10.

Edwards, P. 1986. Duck-fish integrated farming systems. *In* D.J. Farrell & P. Stapleton, eds. *Duck production science and world practice.* Armidale, N.S.W., Australia, University of New England.

Edwards, P. & Kaewpaitoon, K. 1982. Integrated fish farming in Thailand. *ICL ARM Newsl.,* 5(3): 3-4.

Edwards, P., Pullin, R.S.V. & Gartner, J.A. 1988. Research and education for the development of integrated crop-livestock-fish farming systems in the tropics. *ICLARM Stud. Rev.,* (16): 53 p.

Eli, A.B. 1989. Ecological principles of the rice-cum-fish farming system. *In* E.A. Huisman, N. Zonneveld & A.H.M. Bouwmans, eds. *Aquaculture research in Asia: management techniques and nutrition.* Wageningen, Netherlands, Center for Agricultural Publishing and Documentations Pudoc.

Engle, C.R. 1987. Optimal product mix for integrated livestock-fish culture systems in limited resource farms. *J. World Aquacult. Soc.,* 18(3): 137-147.

Eusebio, J.A., Rabino, B.I. & Eusebio, E.G. 1976. Recycling system in integrated plant and animal farming. *Tech. Bull.,* 1(1): 30 p. Laguna, Philippines, NSDB-UPLB Integrated Research Program.

Fang, Y.X., Guo, X.Z., Wang, J.K., Fang, X.Z. & Liu, Z.Y. 1986. Effects of different animal manures on fish farming. *In* J.L. Maclean, L.B. Dizon & L.V. Hosillos, eds. *The First Asian Fisheries Forum*, p. 117-120. Manila, Philippines, Asian Fisheries Society.

Ghosh, A., Chattopadhyay, G.N. & Chakraborty, P.K. 1985. Rice-cum-fish cultivation in coastal paddy fields, package of practices for increasing production. *Aquacult. Ext. Man. New Ser.*, (6): 16 p.

Ghosh, A., Saha, S.K., Banerjee, R.K., Mukherjee, A.B. & Naskar K.R. 1985. Package of practices for increased production in rice-cum-fish farming system. *Aquacult. Ext. Man. Cent. Inland Fish. Res. Inst., Barrackpore (New Ser.)*, (4): 14 p.

Ghosh, S.K. & Pathak, S.C. 1988. Economics of paddy-cum-fish culture in the north eastern states of India. *In* M. Mohan Joseph, ed. *The First Indian Fisheries Forum*, p. 445-449. Mangalore, Karnataka, India, Asian Fisheries Society, Indian Branch.

Ghosh, S.K., Mandal, B.K. & Borthaicur, D.N. 1984. Effects of feeding rates on production of common carp and water quality in paddy-cum-fish culture. *Aquaculture*, 40(2): 97-101.

Halwart, M. 1991. Concurrent rice-fish trials, comparing regular and border planting patterns. *Aquabyte*, 4(1): 2-3.

Han, Y.G. & Ding J.Y. 1986. *Comparative study on the effects of fresh and fermented pig manure on fish yield.* 6 p. Bangkok, NACA.

Hopkins, K.D. 1982. Outstanding yields and profits from livestock-tilapia integrated farming. *ICLARM Newsl.*, 5(3): 13.

Hopkins, K. & Cruz, E.M. 1980. High yield but still questions, three years of animal-fish farming. *ICLARM Newsl.*, 3(4): 12-13.

Hopkins, K.D., Inocencio, P. & Cruz, E.M. 1983. Water quality in integrated livestock-fish ponds. *J. World Maricult. Soc.*, 14: 495- 504.

Hu, B. & Zhou, E. 1990. Review on development of integrated fish farming in China. Wuxi, China, Chinese Academy of Fisheries Sciences, Freshwater Fisheries Research Center.

Hu, B.T. & Yang, H.Z. 1984. *Integrated management of fish-cum-duck farming and its economic efficiency and revenue.* NACA/WP/84/14, 4 p. Bangkok, NACA.

The ICLARM-CLSU integrated animal-fish farming project, poultry-fish and pig-fish trials. 1981. ICLARM Tech. Rep. (2): 29 p.

Jhingran, A.G. 1986. Integrated fish-livestock-crop farming and its role in developing rural economy. *Bull. Cent. Inland Fish. Res. Inst., Barrackpore*, (48): 1.1-4.

Jhingran, V.G. & Sharma, B.K. 1986. Integrated livestock fish farming in India. *In Research plans for integrated aquaculture*, p. 46-53. Dhaka, Bangladesh, FAO/UNDP.

Joel, D.R., Sanjeeviraj, G., Venkataswamy, M. & Natarajan, P. 1985. Prospects for coconut-cum-fish and prawn culture. *Seafood Export J.*, 17(5): 21-25.

Kapur, K. & Lal, K.K. 1986. Relative toxicity of certain livestock wastes for fish culture. *In* J.L. Maclean, L.B. Dizon & L.V. Hosillos, eds. *The First Asian Fisheries Forum.* p. 133-136. Manila, Philippines, Asian Fisheries Society.

Kausar, R. 1983. *Review on recycling of animal wastes as a source of nutrients for freshwater fish culture within an integrated livestock system.* 49 p. Islamabad, Pakistan, Pakistan Agricultural Research Council.

Kuo, S.P. & Ting, Y.Y. 1989. Studies on the water quality of tiger prawn, eel, and fish-cum-duck ponds in southwestern Taiwan. *COA Fish. Ser.*, (16): 113-141.

Lastimoza, P.J. 1982. Spreading the word about rice-fish culture. *PCARRD Monit.*, 10(5): 2-3.

Li, S. 1987. Energy structure and efficiency of a typical Chinese integrated farm. *Aquaculture*, 65(2): 105-118.

Li, K. 1988. Rice-fish culture in China, a review. *Aquaculture*, 71(3): 173-186.

Libuano, L.P. 1989. A goat-fish integrated farming system in the Philippines: effects of stocking densities and goat manure loading rates on the yield of *Oreochromis niloticus*. *In* E.A. Huisman, N. Zonneveld & A.H.M. Bouwmans, eds. *Aquacultural research in Asia: management techniques and nutrition.* Wageningen, Netherlands, Center for Agricultural Publishing and Documentations Pudoc.

Libunao, L.P. 1990. Goat/fish integrated farming in the Philippines. *Ambio*, 19(8): 408-410.

Lightfoot, C. 1990. Integration of aquaculture and agriculture, a route to sustainable farming systems. *Naga, ICLARM Q.*, 13(1): 9-12.

Lightfoot, C. & Tuan, N.A. 1990. Drawing picture of integrated farms helps everyone: an example from Vietnam. *Aquabyte*, 3(2): 5-6.

Lightfoot, C., Dela Cruz, C.R. & Carangal, V.R. 1990. International research collaboration in rice-fish research. *Naga, ICLARM Q.*, 13(4): 10-11.

Lightfoot, C., Roger, P.A., Cagauan, A.G. & Dela Cruz, C.R. 1990. A fish crop may improve rice yields and ricefields. *Naga, ICLARM Q.*, 13(4): 12-13.

Lipton, A.P. 1983. Prospects of paddy-cum-fish culture in the north eastern region of India. *Seafood Export J.*, 15(9): 25-30.

Little, D. & Muir, J. 1987. *A guide to integrated warm water aquaculture.* 238 p. Stirling, Scotland, Institute of Aquaculture, University of Stirling.

Lorico, B.V. 1981. Fish culture integration could maximize use of resources. *PCARRD Monit.*, 9(9):11-12.

Mackay, K.T., Chapman, G., Sollows, J. & Thongpan, N. 1986. Rice-fish culture in North East Thailand, the diversity stability link. 27 p. Paper presented at the IFOAM Sixth International Scientific Conference, California, USA, 18-21 August 1986.

Maclean, J.L. 1987. Small-scale integrated farming in the Philippines. *Naga, ICLARM Q.*, 10(1): 11-12.

Mang-umphan, K. 1987. Culture of Nile tilapia in paddles using chemical and commercial organic fertilizers. *Thai Fish. Gaz.*, 40(2): 135-147.

Mang-umphan, K. & Arce, R.G. 1988. Culture of Nile tilapia *(Oreochromis niloticus)* in a rice-fish culture system using chemical and commercial organic fertilizers. *ICLARM Conf. Proc.*, (15): 59-62.

Manik, R. & Tiensongrusmee, B. 1979. Integrated brackishwater farm system in Indonesia. *Bull. Brackishwat. Aquacult. Dev. Cent. Jepara*, 5(1/2): 369-376.

Middendorp, A.J. & Verreth, J.A.J. 1986. The potential of and constraints to fish culture in integrated farming systems in the Lam Pao Irrigation Project, Northeast Thailand. *Aquaculture*, 56(1): 63-78.

Mitra, B., Gupta A. & Lah, U.K. 1987. Effects of some manures on the growth and production of major carps in village ponds of District Birbhum, West Bengal. *Environ. Ecol.*, 5(2): 381-385.

Mukhopadhyay, P.K., Das, D.N. & Roy, B. 1989. Integrated rice-fish culture, increased productivity from deepwater ricelands in West Bengal. *Naga, ICLARM Q.*, 12(1): 6-7.

Naegel, L. 1988. The integration of aquaculture and agricultural production. *Anim. Res. Dev.*, 27: 7-15.

Natarajan, M. 1985. Recycling of poultry droppings through tilapia fish culture. *Agric. Wastes*, 13(4): 305-313.

Ogburn, D.M., Jumalon, N.A. & Sycip, M.L. 1986. An integrated semi-intensive shrimp and livestock system in the Philippines. *In* J.L. Maclean, L.B. Dizon & L.V. Hosillos, eds. *The First Asian Fisheries Forum*, p. 137-140. Manila, Philippines, Asian Fisheries Society.

Pandit, P.K. & Joshi, H.C. 1986. Control of paddy pests through biological and chemical means in paddy cum fish culture system. *Bull. Cent. Inland Fish. Res. Inst. (Barrackpore)*, (48): 13.1-5.

Paul, S. 1986. Some economic aspects of integrated farming systems. *Bull. Cent. Inland Fish. Res. Inst., Barrackpore*, (48): 19.1-5.

Pinnoi, S. 1986. The economics of integrated fish farm production in the Central Plain, 1985. *Thai Fish. Gaz.*, 39(4): 359-369.

Plavnik, I., Barash, H. & Schroeder, G. 1983. Utilization of duck's droppings in fish farming. *Nutr. Rep. Int.*, 28(3): 635-641.

Polich, T. 1979. Rice/carp farming in the Philippines and cultural acceptance. *ICLARM Newsl.*, 2(4): 13-15.

Potipitak, K., Kribgratok, S. & Sutheemechaiicul, S. 1986. Rice field fish culture in rainfed agricultural development area. *Thai Fish. Gaz.*, 39(6): 575-582.

Pudadera, B.J., Jr., Corre, K.C., Coniza, E. & Taleon, G.A. 1986. Integrated farming of broiler chickens with fish and shrimp in brackishwater ponds. *In* J.L. Maclean, L.B. Dizon & L.V. Hosillos, eds. *The First Asian Fisheries Forum.* p. 141-144. Manila, Philippines, Asian Fisheries Society.

Pullin, R.S.V. 1985. Time to reappraise rice-fish culture. *ICLARM Newsl.*, 8(4): 3-4.

Purushan, K.S. 1986. Recent advances in paddy-cum fish culture and its scope in Kerala. *Seafood Export J.*, 18(5): 13-16.

Quines, O.D., Vargas, R.B. & Ibarra, V.M. 1986. Fecal coliforms as index of pollutions in an integrated pig-fish farm system. *In* J.L. Maclean, L.B. Dizon & L.V. Hosillos, eds. *The First Asian Fisheries Forum.* p. 145-147. Manila, Philippines, Asian Fisheries Society.

Radheyshyam. 1986. Microbiological aspects in integrated fish farming systems. *Bull. Cent. Inland Fish. Res. Inst., Barrackpore,* (48): 17.1-17.

Ramsey, P. 1983. Rice-fish practices in Ifugao Province, Philippines. *ICLARM Newsl.*, 6(3): 8.

Report of the International Workshop on Integrated Livestock/Fish/Crop/ Forestry Production Systems for Small Farmers and Fishermen. Vol. 1. 1980. International Workshop on Integrated Livestock/Fish/ Crop/Forestry Production Systems for Small Farmers and Fishermen, Chiang Mai, Thailand, 4-9 April 1980. Bangkok, Department of Livestock Development.

Roy, B., Das, D.N. & Mukhopadhyay, P.K. 1990. Rice-fish-vegetable integrated farming. towards a sustainable ecosystem. *Naga, ICLARM Q.*, 13(4): 17-18.

Ruddle, K. 1982. Traditional integrated farming systems and rural development, the example of rice field fisheries in Southeast Asia. *Agric. Admin.*, 10: 1-11.

Ruddle, K. & Zhong, G. 1988. Integrated agriculture-aquaculture in South China, the dike-pond system of the Zhujiang Delta. 173 p. Cambridge, Cambridge University Press.

Ruddle, K., Furtado, J.I., Zhong, G.F. & Deng, H.Z. 1983. The mulberry dike-carp pond resource system of the Zhujiang (Pearl River) Delta, People's Republic of China. I. Environmental context and system overview. *Appl. Geogr.*, 3: 45-62.

Ryther, J.H.J. 1983. The evolution of integrated aquaculture systems. *J. World Maricult. Soc.*, 14: 473-484.

Sathiadhas, R., Panlickar, K.K.P., Satyavan, U.K. & Jacob, T. 1989. Economic evaluation of paddy-prawn integrated farming in Kerala. *Seafood Export J.*, 21(11): 9-21.

Shang, Y.C. 1988. Integrated fish farming in China. *Aquacult. Mag.*, 14(2): 28-33.

Shang, Y.C. & Costa-Pierce, B.A. 1983. Integrated aquaculture-agriculture farming systems, some economic aspects. *J. World Maricult. Soc.*, 14: 523-530.

Sharma, B.K. & Das, M.K. 1988. Integrated fish-livestock-crop farming system. *In* M. Mohan Joseph, ed. *The First Indian Fisheries Forum,* p. 27-30. Mangalore, Karnataka, India, Asian Fisheries Society, Indian Branch.

Sharma,B.K., Das, M.K. & Chakrabarty, D.P. 1985. Fish-cum-livestock farming, package of practices for increasing production. *Aquacult. Ext. Mat. New Ser.*, (5): 32 p.

Sharma, B.K., Das, M.K. & Chakraborty, D.P. 1985. Package of practices for increasing production in fish-cum-livestock farming system. *Aquacult. Ext. Man. Cent. Inland Fish. Res. Inst. Barrackpore (New Ser),* (5): 32 p.

Sharma, B.K., Das M.K. & Das, S.R. 1988. Carp farming integrated with pig raising in India. *Environ. Ecol.*, 6(1): 159-168.

Singh, H. 1990. Mixed rice-fish farming in India. *Seafood Export J.*, 22(5): 19-22.

Sinha, V.R.P. 1986. Integrated carp farming in Asian country. *In* R. Billard & J. Marcel, eds. *Aquaculture of cyprinids (L'aquaculture des cyprinids),* p. 377-390. Coll. Hydrobiol. Aquacult.

Sollows, J.D. & Tongpan, N. 1986. Comparative economics of rice-fish culture and rice monoculture in Ubon Province, Northeast Thailand. *In* J.L. Maclean, L.B. Dixon & L.V. Hosillos, eds. *The First Asian Fisheries Forum,* p. 149-152. Manila, Philippines, Asian Fisheries Society.

Symposium-Workshop on Integrated Crop-Livestock-Fish Farming, 1980. Integrated crop-livestock-fish farming. Los Baños, Laguna, Philippines, 19-24 November 1979. ASPAC-FFTC Book Ser. No. 16, 147 p. Taipei, Taiwan, Food and Fertilizer Technology Center, ASPAC.

Tamse, A.F., Fortes, N.R., Catedrilla, L.C. & Yuseco, J.E.H. 1985. The effect of using piggery wastes in brackishwater fishpond on fish production. *Univ. Philipp. Visayas Fish. J.*, 1(1): 69-76.

Taylor, S. & Klampratum, D. 1986. Appropriate adaptation of rice-fish culture to upland terraced rice in Northern Thailand. Paper presented at the Department of Agriculture Annual Conference, Bartok, Thailand, 23-28 April 1986.

Taylor, S.R., Pakdee, B., & Klampratum, D. 1988. Border method and fish culture, synergistic effects on the yield of rice grain. *ICLARM Conf. Proc.*, (15): 91-98.

Tetangko, M.H., editor. 1980. *Integrated crop-livestock-fish farming. Proceedings of the Symposium-Workshop on Integrated Crop-Livestock-Fish Farming,* Los Baños, Laguna, Philippines, 19-24 November 1979. ASPAC-FFTC Book Ser. No. 16, 147 p. Taipei, Taiwan, Food and Fertilizer Technology Center, ASPAC.

Theincharoen, P. 1987. Rice field fish culture in Surin province. *Thai Fish. Gaz.*, 40(3): 265-269.

Udeni, E. 1988. Duck-fish integration under extensive fish culture, without the use of a fish feed. *In* P. Radhakrishna, M.M. Singh & C.K. John, eds. *Second Proceedings of the Asian Conference on Technology for Rural Development.* Singapore, World Scientific Publishing Co.

Wohlfarth, G.W. & Schroeder, G.L. 1979. Use of manure in fish farming, a review. *Agric. Wastes,* 1(4): 279-299.

Yan, J.H. 1989. Integrated fish culture management in China. *In* W.J. Mitch & S.E. Joergensen, eds. *Ecological engineering: an introduction to ecotechnology,* p. 375-408. New York, USA, John Wiley & Sons.

Zhu, Y., Yang, Y., Wan, J., Hua, D. & Mathias, J.A. 1990. The effect of manure application rate and frequency upon fish yield in integrated fish farm ponds. *Aquaculture,* 91(3-4): 233-251.

List of Participants*

BANGLADESH

Dr. Mahfuzuddin Ahmed
Technical Coordinator
Bangladesh Fish Culture Extension Impact Study Project
International Center for Living Aquatic Resources Management

Dr. Modadugu V. Gupta
Aquaculture Specialist
International Center for Living Aquatic Resources Management - Bangladesh

Mr. Golam Azam Khan
Senior Assistant Director
Department of Fisheries

Mr. Francisco Noble
Fisheries Expert
Association of Development Agencies in Bangladesh

Dr. Eric Worby
Anthropologist
International Center for Living Aquatic Resources Management - Bangladesh

CANADA

Mr. John Sollows
Adviser/Biologist
Agro-Dev Canada Inc.

CHINA, PEOPLE'S REPUBLIC OF

Prof. Yixian Guo
Institute of Crop Breeding and Cultivation
Chinese Academy of Agricultural Sciences

Prof. Bao Tong Hu
Associate Professor/Director
Department of Fisheries-Economics
Asia-Pacific Regional Research and Training Center for Integrated Fish Farming

Mr. Kuanhong Min
Trainer/Lecturer
Asia-Pacific Regional Research and Training Center for Integrated Fish Farming

Prof. Huazhu Yang
Associate Professor/Deputy Director
Department of Integrated Fish Farming
Asia-Pacific Regional Research and Training Center for Integrated Fish Farming

GHANA

Dr. Mark Preir
Associate Scientist
International Center for Living Aquatic Resources Management - Ghana Aquaculture Project
Institute of Aquatic Biology

INDIA

Mr. Bal Krishan Sharma
Principal Scientist/Chief Training Organizer
Central Institute of Freshwater Aquaculture

Dr. Satyendra D. Tripathi
Director
Central Institute of Freshwater Aquaculture

MALAWI

Dr. Reg Noble
Research Scientist
International Center for Living Aquatic Resources Management - Africa Aquaculture Project

MALAYSIA

Mr. Raihan Sh. Hj. Ahmad
Officer
Department of Fisheries

Dr. Ahyaudin B. Ali
Associate Professor, Aquatic Biology
School of Biological Sciences
Universiti Sains Malaysia

PHILIPPINES

Ms. Mary Ann P. Bimbao
Program Assistant/Economist
International Center for Living Aquatic Resources Management

Mr. Jens Peter Tang Dalsgaard
Assistant Scientist
International Center for Living Aquatic Resources Management

Dr. Catalino dela Cnuz
Senior Specialist
International Center for Living Aquatic Resources Management

Mr. Franklin V. Fermin
Aquaculture Specialist
International Institute of Rural Reconstruction

Mr. Matthias Halwart
Collaborative Research Scientist
Asian Rice Farming Systems Network
International Rice Research Institute

Prof. Ruben Sevilleja
Director
Freshwater Aquaculture Center
Central Luzon State University

Mr. Jose Torres
Senior Agriculturist
Department of Agriculture

SENEGAL

Dr. Anis Diallo
Aquaculturist
Centre de Recherches Oceanograhiques Dakar-Thiaroye

THAILAND

Mr. David Little
Assistant Professor
Division of Agricultural and Food Engineering
Asian Institute of Technology

VIETNAM

Mr. Le Thanh Duong
Deputy Director
Faming Systems Research and Research Development Center
University of Cantho

Mr. Le Thanh Hung
Faculty of Fisheries
University of Agriculture and Forestry

*Designations and affiliations of participants were as of February 1992, at the time of the workshop.

WORKSHOP SUPPORT STAFF

Technical Specialists
Dr. Julian F. Gonsalves
Mr. Scott A. Killough
Dr. Clive Lightfoot
Dr. Roger S.V. Pullin
Mr. Jaime P. Ronquillo

Communication Specialist
Ms. Ines Vivian D. Domingo

Editors
Ms. Lyn Capistrano-Doren
Ms. Noreen O. Luna

Artists
Mr. Florante C. Belardo
Mr. Ricardo E. Cantada
Mr. Albert Contemprate
Mr. Mitchell Doren
Mr. Bernabe Remoquillo
Mr. Roger M. Villar

Administrative and Secretarial Support
Ms. Gina Marie M. Bautista
Ms. Estella S. Kasala
Ms. Alelie D. Miranda
Ms. Angelita T. Poblete